OFFSHORE SUPPORT VESSELS

A PRACTICAL GUIDE

by

Gary Ritchie
Master Mariner, BA(Hons) MNI

OFFSHORE SUPPORT VESSELS
A PRACTICAL GUIDE

by Gary Ritchie
Master Mariner, BA(Hons) MNI

Published by The Nautical Institute
202 Lambeth Road, London SE1 7LQ, England
Telephone +44 (0)207 928 1351   Fax +44 (0)207 401 28127
website: www.nautinst.org

First edition published 2008
Copyright © The Nautical Institute 2008

All rights reserved. No part of this publication may be reproduced, stored in a retrieval system, or transmitted in any form, by any means, electronic, mechanical, photocopying, recording or otherwise, without the prior written permission of the publisher, except for the quotation of brief passages in reviews. Although great care has been taken with the writing of the book and production of the volume, neither The Nautical Institute nor the author can accept any responsibility for errors or omissions or their consequences.

The book has been prepared to address the subject of multi-purpose offshore support vessels. This should not, however, be taken to mean that this document deals comprehensively with all of the concerns that will need to be addressed or even, where a particular matter is addressed, that this document sets out the only definitive view for all situations. The opinions expressed are those of the author only and are not necessarily to be taken as the policies or views of any organisation with which he has any connection.

Readers should make themselves aware of any local, national or international changes to bylaws, legislation, statutory and administrative requirements that have been introduced which might affect any decisions taken on board.

Cover picture courtesy of Huisman Itec

Typeset by J A Hepworth FNI
1 Ropers Court, Lavenham, Suffolk CO10 9PU, England
www.hepworth-computer-services.co.uk

Printed in England by Modern Colour Solutions
2 BullsBridge Ind Est, Hayes Road,
Southall, Middlesex UB2 5NB, England

ISBN 1 870077 88 1

FOREWORD 1
by Mr S. A. McNeill CEng MRINA
Vice President Vessels and Equipment Subsea 7

As the development of oil and gas fields heads in to ever deeper water and more hostile environments, there has been a dramatic rise in the demand for new vessels, equipment and technology capable of operating in this dramatic arena. The scale, complexity and innovation of the solutions required to solve the constantly evolving challenges of deepwater offshore construction are rarely paralleled in human history. Offshore Support Vessels operate in the harshest working conditions known to man and their reliability is of paramount importance to protect their crews, the environment and the infrastructure of the fields where they operate.

In such an exciting and complex field of engineering there is very limited reference material and so I was very pleased to read Gary Ritchie's new book that seeks to advance the reader's understanding of this subject. At a time when there is an unprecedented boom in the construction of new Offshore Support Vessels, with a corresponding peak in the demand for new personnel to join the industry, such a book could hardly be more timely. For the student or newcomer to the offshore subsea construction industry this book provides an excellent starting point and an invaluable introduction to the principles, equipment and vessels utilised in the offshore subsea construction environment. For those already in the industry, who may have learnt the hard way, this book may also serve them well if they are seeking to expand their knowledge. For other professionals, who need to gain an understanding of the offshore construction industry and its unique terminology, this is a valuable guidebook.

This book focuses on the basic principles of some of the cornerstone technologies in offshore subsea construction and should be of assistance to those either planning operations or specifying and designing new offshore support vessels. Topics covered include Dynamic Positioning Systems, Pipelay Vessel Operations, Remotely Operated Vehicles and their host support vessels. In addition Saturation Diving Vessels, Diving Equipment and Diving Operations are discussed at length. Saturation Diving is one of the original foundations of subsea construction activities and is still very relevant today, since, at the time of writing, more new diving vessels are under construction than at any time in the past. Finally, but most importantly, the author introduces some of the key international legislation and principles governing the safe, secure and environmental operation of Offshore Support Vessels.

There are many different solutions to the numerous novel problems posed to Offshore Vessel Operators as we advance into this frontier science. It is essential that all are carefully assessed, making best use of sound engineering principles, to ensure we make the right choices to protect ourselves and the environment as we sail cautiously forward. I believe that this book is a valuable contribution and hope it will provide you with a good headstart on your journey into the deep.

FOREWORD 2

I have known Gary Ritchie as a work colleague for more than a decade now and have always admired his dedication to his craft, his great attention to detail and his thorough knowledge of offshore support vessel operations gained in a great part from practical experience. It is this great attention to detail and thorough knowledge which leads me to believe that there is no one better suited to compile a book on offshore support vessel operations.

The offshore oil and gas operators depend upon various different types of offshore vessels to provide the support required to enable their particular workscopes to be performed. Although in the past there have been books written on certain types of offshore support vessels, this is the first time that such a comprehensive book covering many various different types of offshore support vessel operations has been available. In fact, there is no other book in existence which contains such a comprehensive text on up to date offshore support vessel operations.

I would recommend this book to both offshore and onshore personnel involved in offshore support vessel operations including those who are new to the industry together with those who are new to certain vessel types. This practical guide to Multi-Purpose offshore support vessels is a very valuable book containing information equally valid for offshore personnel as well as others involved in the industry and can provide valuable reference to onshore management enabling them to gain an insight into particular issues offshore.

Captain Gary McKenzie
Master Mariner, MNI, MIOSH, M.Inst.Pet
October 2007

INTRODUCTION

The Offshore Industry is a varied sector within which many vessel types operate, performing numerous different tasks with often unique systems and equipment. These vessels can range from purpose built specialised ships which may, for example, only perform diving operations, to vessels which have been repeatedly converted from one vessel type to another as the nature of the business changes. As such, the subject of Multi-Purpose Offshore Support Vessels covers a very broad spectrum of vessel types and vessel operations and it is therefore very difficult to provide a definitive overview of the subject matter.

However, there are many standard features, systems and operating practices that are applicable across the industry. It is these generic features that this book therefore proposes to introduce, whilst particular reference is made to a number of specific vessels in order to illustrate the diversity and complexity of the systems involved.

It is hoped that by presenting a general overview and introduction to Dive Support, ROV Support, Construction Support and Pipe Lay Vessels, the text will provide an insight to this specialised sector, not only for anyone planning to transfer or commence a career within the industry, but for those already established within such a diverse business.

ACKNOWLEDGEMENTS

The author wishes to thank the following who kindly provided valued assistance in the preparation of this book: Captain Gary McKenzie, Steph McNeil, Dave Dobeson, Jackie Doyle, Elaine Percival, Alex Main, Allan Cameron, Paul McBurnie, Denis Johnstone, John Patterson, Chris Fletcher, Derek Gray, Bruce McKenzie and Bob Barr for their valuable assistance and comments.

Keith Phillips at Guidance Navigation Limited for providing details and images relating to the RadaScan DP Reference System. These are reproduced by kind permission.

Mark Prise of Scotgrip (UK) Limited for providing details of Scotgrip safety products. Images in section 13 are reproduced by kind permission.

Mark Williams, Human Element Development Manager, Maritime and Coastguard Agency. The images in section 13 are reproduced with permission of the Maritime and Coastguard Agency. All material remains the worldwide copyright of MCA / Crown and may not be reproduced without written permission.

Hugh Williams of the International Marine Contractors Association (IMCA) for allowing permission to use the IMCA Safety Posters 'Manual Handling' and 'Slips, Trips and Falls'. The posters are reproduced here under kind permission, courtesy of IMCA.

Figure 1.6 is reproduced by kind permission of Fuglefjellet.

The majority of the photographs contained within this book are reproduced by kind permission of Subsea 7 from their extensive database. All due care has been taken to only include photographs available from this database and any failure to correctly identify the copyright holder is unintentional.

Thank You Whoever You Are (h, Steve, Mark, Pete and Ian).

For Linda, Ewan and Calum.

CONTENTS

	page
Foreword 1	iii
Foreword 2	iv
Introduction	v
Acknowledgements	vi
Contents	vii
List of Figures	viii

Chapter   **page**

1	Offshore Support Vessel Design	1
2	Offshore Support Vessel Design — Dive Support Vessels	15
3	Offshore Support Vessel Design — ROV Support Vessels	37
4	Offshore Support Vessel Design — Construction Vessels	47
5	Offshore Support Vessel Design — Pipe Lay Vessels	53
6	Diving Operations	65
7	ROV Support Vessel Operations	73
8	Construction Operations	81
9	Pipe Lay Operations	89
10	Dynamic Positioning Systems	95
11	ISM Code	107
12	ISPS Code	113
13	Shipboard Safety	117
14	Environmental Management	145
15	Ballast Management	151

Appendix

A	A Brief History of Saturation Diving Systems and ROVs	155

Index	157

LIST OF FIGURES

Figure		page
1.1	*Rockwater 1* — Dive Support Vessel	1
1.2	*DSND Pelican* — Dive Support Vessel	1
1.3	*Toisa Polaris* — Dive Support Vessel	2
1.4	*Kommandor Subsea* — ROV Support Vessel	2
1.5	*Seisranger* — ROV Support Vessel	3
1.6	*Normand Seven* — ROV Support Vessel	3
1.7	*Subsea Viking* — Multi-Purpose Offshore Support Vessel	3
1.8	*Subsea Viking* — Multi-Purpose Offshore Support Vessel	4
1.9	*Toisa Polaris* — Dive Support Vessel with Construction Capabilities	5
1.10	*Subsea Viking* — 100 tonne SWL Huisman Crane	5
1.11	*Skandi Navica* — Pipe Lay Vessel	6
1.12	*Seven Oceans* — Pipe Lay Vessel	6
1.13	Tunnel Thruster (with guards)	6
1.14	Stern Azimuth Thrusters	7
1.15	Retractable Stern Azimuth Thruster	7
1.16	*Seisranger* — Main Propeller	8
1.17	Forward Bridge Console	9
1.18	*Subsea Viking* — Forward Bridge Console for Navigation and Transit Control	10
1.19	*Subsea Viking* — Dining Facilities	11
1.20	ROV Moonpool	11
1.21	Aft Working Moonpool fitted with Multi-Purpose Handling System (MPHS)	12
1.22	Upper and Lower Hatch Covers for Moonpool	12
1.23	A-Frame Arrangement	12
1.24	A-Frame Arrangement	12
1.25	Offshore Support Vessel Helideck	13
1.26	UK Helideck Markings	13
1.27	Helicopter Landing Offshore	14
2.1	*Rockwater 2* — Dive Support Vessel	15
2.2	Diver and Diving Bell	15
2.3	Dive System Class Notations — Det Norske Veritas	16
2.4	Environmental Limits for Monohull Vessels — Det Norske Veritas	16
2.5	Environmental Limits for Semi-Submersible Vessels — Det Norske Veritas	16
2.6	*Toisa Polaris* — Class III Dive Support Vessel	17
2.7	Dive System Chamber Complex	18
2.8	Dive System Chamber Complex	18
2.9	Dive System Chamber and Bell Complex	19
2.10	Living Chamber — Equipment Lock	19
2.11	Living Chamber — Internal	20
2.12	Storage of Gas Cylinders	21
2.13	Marking of Gas Cylinders	21
2.14	Built-in Breathing System (BIBS) face mask	22

Figure		page
2.15	Bell Mating Trunking and Clamp	24
2.16	Saturation Diving Bell	25
2.17	Internal Diving Bell Control Panels	25
2.18	Diving Bell Main Umbilical Cross-section	26
2.19	Diving Bell Main Umbilical Winch	26
2.20	Diver's Umbilical	27
2.21	Saturation Diving Bell	29
2.22	Schematic of Diving Bell, Moonpool Cursor and Overhead Trolley Arrangement	30
2.23	Schematic of Diving Bell, Moonpool Cursor and Overhead Trolley Arrangement	30
2.24	Guide Wire Weight	30
2.25	Saturation Control Room	32
2.26	Dive Control Room	32
2.27	Self-Propelled Hyperbaric Lifeboat	33
2.28	Hyperbaric Lifeboat Mating	33
2.29	Hyperbaric Lifeboat — External Marking	34
2.30	Hyperbaric Lifeboat — Lifting Beam	34
2.31	Hyperbaric Evacuation Trunk Access from Living Chamber	35
2.32	Diver's Personal Equipment — Helmet	36
3.1	*Kommandor Subsea* — ROV Support Vessel	37
3.2	Autonomous Underwater Vehicle	38
3.3	Hercules Workclass ROV	39
3.4	Centurion HD Workclass ROV	39
3.5	Seaeye Tiger Observation Class ROV	39
3.6	Workclass ROV Frame	40
3.7	Buoyancy Modules	40
3.8	Centurion ROV — Thruster Units	41
3.9	ROV Unit with Manipulators	42
3.10	Titan 4 Manipulator	42
3.11	Tooling Skid	43
3.12	ROV and Tooling Skid	43
3.13	ROV Deployment and Main Umbilical	43
3.14	Observation Class ROV with TMS Garage	43
3.15	ROV Control Station	44
3.16	Example ROV Pilot Console Display	44
3.17	Example ROV Pilot Console Display	44
3.18	Moonpool Launch and Recovery System	45
3.19	Umbilical Winch	45
3.20	ROV Umbilical — Armoured Exterior	45
3.21	Overside ROV Launch and Recovery System	46
3.22	Overside ROV Launch and Recovery System	46
3.23	Overside A-Frame Launch and Recovery System	46

Figure		page
4.1	Boom Type Crane with Lattice Arrangement	47
4.2	Lattice Type Boom Crane with Luffing Wires	48
4.3	Lattice Type Boom Crane with Luffing Wires	48
4.4	Crane Capacity Curves in Tabular Format	49
4.5	Crane Capacity Curves in Graphical Format	50
4.6	Lattice Boom Type Crane	51
4.7	Box Boom Type Crane with Luffing Cylinders	51
4.8	Knuckle Boom Crane	51
4.9	Knuckle Boom Crane in Stowed Position	52
4.10	Ballast transfer and stability management are essential components of crane vessel operations	52
5.1	*Skandi Navica* — Pipe Lay Vessel	53
5.2	Reel Lay and Ramp Arrangement	55
5.3	*Skandi Navica* — Deck Reel and Stern Ramp Deployment Arrangement	56
5.4	Ramp Arrangement	56
5.5	Ramp Arrangement	56
5.6	Deck Mounted Reel — External	57
5.7	Deck Mounted Reel — Internal	57
5.8	Deck Mounted Reel Capacities	57
5.9	Deck Mounted Reel — Tie-in Arrangement	57
5.10	Ramp Arrangement	57
5.11	Main and Piggy Back Aligners	58
5.12	Main Aligner Track	58
5.13	Piggy Back Straightener	58
5.14	Tracked Tensioner Arrangement	59
5.15	Pipe Clamp Arrangement	59
5.16	Exit Monitoring Frame	59
5.17	Exit Roller Box	60
5.18	Abandonment and Recovery Sheave at the Base of the Tensioners	60
5.19	PLET Handling Frame for Deployment of an End Termination	60
5.20	*Toisa Perseus* — Vertical Lay Tower and Under Deck Carousels	60
5.21	Under Deck Reel Capacities	61
5.22	Under Deck Carousel Spooler and Carousel	61
5.23	Loading Product onto Under Deck Carousel	61
5.24	Deck Mounted Carousel	61
5.25	Minimum and Maximum Squeeze Pressure Arrangement on Loading Tower	61
5.26	Loading Tower and Deck Radius Controller	62
5.27	Loading Tower Tensioner Track	62
5.28	Deck Radius Controller	62
5.29	Deck Radius Controller	62
5.30	Vertical Lay System Tower — Upper Chute	63
5.31	Track Tensioners	63
5.32	Vertical Lay System Tower with Track Tensioners	63
5.33	Vertical Lay System Tower with Track Tensioners	64

Figure		page
6.1	Preparation for Diving Operations	66
6.2	Diver's Excursion Umbilical	69
6.3	Diver at Work Subsea	69
6.4	Diver exiting the Dive Bell	70
7.1	*Kommandor Subsea 2000* — ROV Support Vessel	73
7.2	ROV Launch and Recovery	73
7.3	ROV Integrity Checks	76
8.1	Offshore Crane Operations	81
8.2	Wire Arrangement — Single Bend	85
8.3	Wire Arrangement — Double Bend	85
8.4	Wire Arrangement — Continuous Bend	85
8.5	Wire Arrangement — Reverse Bend	85
8.6	Knuckle Boom Crane — Sheave Arrangement	86
8.7	Incorrect Spooling can lead to Slack Turns	87
9.1	Pipe Lay Operations — Stern Ramp Lay System	89
9.2	*Seven Oceans* — Pipe Lay Vessel	89
9.3	Pipe Lay Spooling Operations	91
9.4	Pipe Lay Loading Operations — Securing the Pipe End	92
9.5	Pipe Lay Loading Operations — Deck Mounted Reel	92
9.6	Pipe Lay Deployment Operations	93
9.7	*Skandi Navica* — Pipe Lay Vessel	94
10.1	Forces involved in Dynamic Positioning	95
10.2	Typical Dynamic Positioning System	95
10.3	Differential Global Positioning System (DGPS)	96
10.4	Transponder Beacons	97
10.5	Fanbeam Laser Unit System	97
10.6	Taut Wire Arrangement	98
10.7	RadaScan Transponder	99
10.8	RadaScan Sensor	99
10.9	RadaScan 170° Diagram	99
10.10	RadaScan 360° Diagram	99
10.11	DP Classification (DNV and NMD)	101
10.12	DP Classification (Lloyds Register)	101
10.13	DP Classification — Comparison Summary	101
10.14	IMCA Recommended Experience Levels — Existing Vessels	103
10.15	IMCA Recommended Experience Levels — New Vessels	103
10.16	IMCA Organisation	104
11.1	Example Safety and Environmental Policy	109

Figure		page
11.2	*Olympic Canyon* — ROV Support Vessel	111
12.1	*Subsea Viking* — Offshore Support Vessel	115
13.1	Principles of Health and Safety	119
13.2	Human Factor Statistics	120
13.3	Shipboard Safety Organisation	128
13.4	Dive Support Vessel — Safety Inspection Areas	130
13.5	*Skandi Navica* — Pipe Lay Vessel	131
13.6	*Toisa Polaris* — Dive Support Vessel	133
13.7a	Gangway Access	133
13.7b	Accommodation Ladder Access	134
13.8	Pipe and Cable Bridges	134
13.9	Round Bar Rung Covers	134
13.10	Kicker Treads on External Stairways	135
13.11	IMCA Slips, Trips and Falls Poster	135
13.12	Pilot Boarding Operations	136
13.13	Risk Analysis Matrix	137
13.14	Mooring Ropes	140
13.15	Mooring Ropes	141
13.16	IMCA Manual Handling Poster	142
13.17	*Toisa Perseus* — Pipe Lay Vessel	143
14.1	Summary of Garbage Disposal	146
14.2	Controlled and Hazardous Waste	148
15.1	Harmful Aquatic Organisms and Pathogens	151

Chapter 1

OFFSHORE SUPPORT VESSEL DESIGN

General Introduction

In order to detail the main design characteristics which may be expected on an Offshore Support Vessel, it is important to define the vessel types that are included within this category and therefore to define the roles that these vessels may be required to fulfil. The term Offshore Support Vessel can include many vessel types and it is unusual for one single vessel to only fulfil one particular function, therefore one vessel, such as the *Rockwater 1*, can perform diving, ROV, survey and construction support operations.

The following introductions to the various functions that can be performed by the Offshore Support Vessel are therefore provided as a general indication. More detailed role specific design features will be examined further in the subsequent sections.

Dive Support Vessels

Dive Support Vessels within the offshore industry can range from converted vessels fitted with rudimentary air diving spreads to purpose built vessels fitted with extensive and complex saturation diving systems. The equipment and systems required for an air diving operation is obviously vastly different to that required for deepwater saturation diving operations and there are many differences in design and operation from one system to another.

However, the general design principles for a Dive Support Vessel will include those for a generic Offshore Support Vessel, with additional requirements for life support and chamber systems, diving bell, diving bell handling systems and emergency evacuation systems.

Figure 1.2 DSND Pelican – Dive Support Vessel

Generic design features for Dive Support Vessels can be summarised as follows:

- A high level of position accuracy and excellent station keeping capabilities are essential for any vessel from which diving operations will be performed. Dive Support Vessels are therefore fitted with fully redundant Dynamic Positioning Systems. Dive Support Vessels are therefore expected to be as a minimum DP Class 2 – DYNPOS AUTR. This requires the system to be provided with redundancy in technical design and with an independent joystick back-up. For certain operations, DP Class 3 – DYNPOS AUTRO will be required. This requires the system to be provided with redundancy in technical design and with an independent joystick

Figure 1.1 Rockwater 1 – Dive Support Vessel

back-up. In addition, DP Class 3 requires a back-up dynamic positioning control system in an emergency dynamic positioning control centre, designed with physical separation for components that provide redundancy.

- Manoeuvring and propulsion systems are of major importance with the requirement to operate the vessel safely and effectively at either slow speed or whilst static and to maintain the vessel's position to a very high degree of accuracy. Redundancy of machinery, manoeuvring and propulsion systems is of paramount importance due to the potential to cause injury to divers if any loss of position occurs.

- The location and type of thrusters and main propulsion utilised is determined to ensure the safety of the divers in appropriate water depths and for the systems maintained onboard the vessel. Generally the diving systems will be located the maximum distance horizontally from any thrusters or propulsion units.

- A protected and stabilised location for the diving chambers, bells and bell handling systems is essential. Generally the chambers and bells will be positioned along the centreline of the vessel with the living chambers being well protected by the ships structure. Access to the self-propelled hyperbaric lifeboat is taken into consideration and launching of the bells through a centreline moonpool offers protection and a stabilised platform for diving operations, protected from the full effects of the weather conditions.

- Dual bridge set-ups are standard for Dive Support Vessels with all main and auxiliary controls being duplicated on forward facing and after facing consoles. Although it is unlikely that the bridge officers will be able to see the bell launching systems from the bridge, a good overall visibility of the working deck and surrounding area of operation is essential. This is of particular importance when multiple operations may be being performed consecutively, such as diving operations and crane operations. A full appreciation of the position of the crane, crane wire and hook, whilst divers are in the water, is therefore critical.

ROV Support Vessels

ROV Support Vessels can include vessels fitted with portable launching systems which can be mobilised and demobilised to the vessel within very short periods of time and form the most rudimentary ROV support systems. Advanced systems permanently fitted to purpose built ROV Support Vessels can include moonpool launched heave compensated handling systems and auxiliary side launched systems.

Figure 1.4 Kommandor Subsea - ROV Support Vessel

As with Dive Support Vessels, a number of design features of systems and equipment onboard ROV Support Vessels can differ from vessel to vessel. However, as all these vessels will be performing similar functions, a number of common or generic design characteristics can be summarised as follows:

- A high level of position accuracy and excellent station keeping capabilities are essential for any vessel from which ROV operations will be performed. The potential for injury to personnel is significantly less with ROV operations compared with diving operations and therefore the requirements for Dynamic Positioning System redundancy is reduced and DP Class 1 or DP Class 2 can be acceptable dependent on the operation proposed.

- Closely linked to the Dynamic Positioning System, manoeuvring and propulsion systems are of major importance with the requirement to operate the

Figure 1.3 Toisa Polaris - Dive Support Vessel

vessel safely and effectively at either slow speed or whilst static and to maintain the vessel's position to a very high degree of accuracy. Redundancy of machinery and manoeuvring and propulsion systems is of paramount importance due to the potential to cause damage to equipment or assets if any loss of position occurs. The location and type of thrusters and main propulsion utilised is determined to ensure the integrity of the equipment in appropriate water depths and for the systems maintained onboard the vessel. Generally the ROV systems will be located at the maximum distance horizontally from any thrusters or propulsion units.

- A protected and stabilised location for the ROV handling systems is essential. Generally the ROVs will be positioned along the centreline of the vessel. Launching of the ROVs through a centreline moonpool offers protection and a stabilised platform for ROV operations, protected from the full effects of the weather conditions.

Figure 1.7
Subsea Viking - Multi-Purpose Offshore Support Vessel

Figure 1.5   Seisranger - ROV Support Vessel

Figure 1.6   Normand Seven - ROV Support Vessel

OFFSHORE SUPPORT VESSELS 3

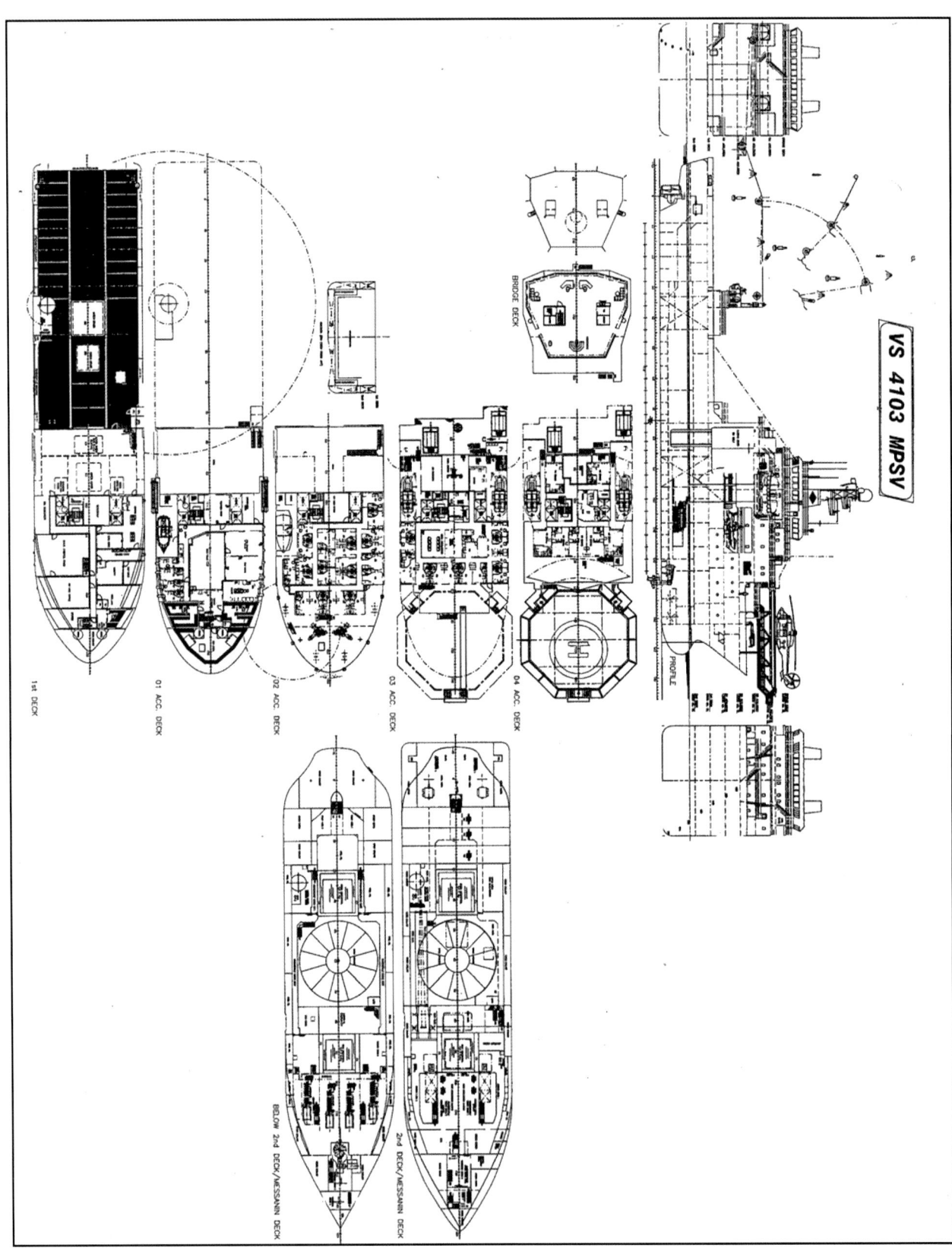

Figure 1.8 Subsea Viking - Multi-Purpose Offshore Support Vessel

Figure 1.9
Toisa Polaris - Dive Support Vessel with Construction Capabilities

- A dual bridge set-up is standard for ROV Support Vessels with all main and auxiliary controls being duplicated on forward facing and after facing consoles. This is of particular importance for Dynamic Positioning and communications controls. Although it is unlikely that the bridge officers will be able to see the ROV launching systems from the bridge, a good overall visibility of the working deck and surrounding operations is essential. This is of particular importance when multiple operations may be being performed consecutively, such as ROV operations and crane operations. A full appreciation of the position of the crane, crane wire and hook, whilst ROVs are in the water, is therefore critical.

Figure 1.10 Subsea Viking - 100 tonne SWL Huisman Crane

Construction Vessels

Offshore Construction Vessels will have many of the generic characteristics and design features associated with Dive and ROV Support Vessels, with the following further considerations:

- A high level of position accuracy and excellent station keeping capabilities are essential during operations where any subsea or surface load is being transferred from or to the offshore construction vessel's deck. In such circumstances the potential for injury to personnel or damage to the vessel or the load due to uncontrolled actions is to be kept at a minimum. The stable maintenance of the vessel's position is therefore essential.

- The main function of the Offshore Construction Vessel will be for the installation and decommissioning of subsea and surface structures and installations, therefore the type, capacity and positioning of the crane will be one of the main design considerations.

- The positioning of the crane will be determined to provide the maximum outreach on the preferred side of the vessel with the maximum capacity at the optimum outreach. The crane driver should be provided with a good overview of all deck and overside areas with no obstructions present to obscure his view.

- Dedicated ballast or counter weight systems will be in place for use in heavy lift operations, where the stability of the vessel will be affected by the weight and position of any loads being loaded or discharged.

- Heave compensation may be provided for the crane to allow for safe and accurate load positioning operations taking into consideration the prevailing sea conditions.

- In order to assist with the loading and discharging of lifts, dedicated tugger winches may be provided. Such crane tugger winches should be provided with dedicated and integrated control systems which are controllable by the crane operator. Generally such devices will be provided with foot pedal controls to allow the crane operator to operate the tuggers whilst using the hand operated main crane controls.

Pipe Lay Vessels

A number of design features of systems and equipment onboard Pipe Lay Vessels can differ from vessel to vessel. However, as all these vessels will be performing similar functions, a number of common or generic design characteristics can be summarised below:

- The main function of the Pipe Lay Vessel will be to

Figure 1.11   Skandi Navica - Pipe Lay Vessel

lay pipe along a designated seabed channel or route and as such the accuracy of the vessel's position keeping capabilities whilst the vessel is moving slowly along this intended channel or route is the most important aspect of the vessel's operation. As such the vessel type will be fitted with DP reference systems conducive to continued movement, such as DGPS and acoustic systems. The ability to interface the pipe lay operations effect on tensions (in the pipe being laid) with the vessel's position keeping abilities is of particular importance.

- In order to maintain the vessel's position along a specified channel or route, the vessel's manoeuvring and propulsion systems are of major importance. The location and type of thrusters and main propulsion utilised is determined to ensure that the possible contact between the pipe lay system and the vessel's manoeuvring systems is minimised.
- Storage of the product, whether rigid or flexible pipe, will be such as to ensure a smooth and unobstructed loading and deployment methodology in port and at sea. Many systems such as those onboard the *Toisa Perseus* and the *Subsea Viking* consist of underdeck storage carousels, whilst the system onboard the *Skandi Navica* has an on deck reel system. The loading and deployment system provided will ensure that maximum pipe bending radius requirements are not exceeded throughout the operations.
- As with the majority of Offshore Support Vessels, a dual forward and after control station will often be in place for DP and manual manoeuvring operations. For pipe lay operations, the vessel will be steaming at slow speed. As such, an ability to control the vessel while visually observing the after deck and pipe lay operations and system is essential for the safe operation of the vessel.

Manoeuvring Systems

All aspects of Offshore Support Vessels' work scope, whether it involves divers, remotely operated vehicles, survey operations, crane operations or flexible or rigid pipe lay operations, requires the vessel to remain in as stable and as accurate a position as is possible. As such, the manoeuvring systems onboard any Offshore Support Vessel are required to be numerous, powerful and highly efficient. On vessels such as Dive Support Vessels, the requirement to maintain the vessel in a highly accurate position whilst human diver intervention operations take place, requires not only a variety of manoeuvring systems to be in place, but also requires a level of full redundancy in case of system failure or blackout. As such the manoeuvring systems onboard such vessels is far more important in the design and construction of the vessel than in other vessel types. The variety of manoeuvring systems available is quite diverse and the various configurations of thruster types, rudders and propellers that can be utilised is such that it is difficult to describe all the available types. However, the following information is provided to show a general overview of the systems currently in use onboard Offshore Support Vessels and the types of systems that may be encountered within the industry.

Tunnel Thrusters

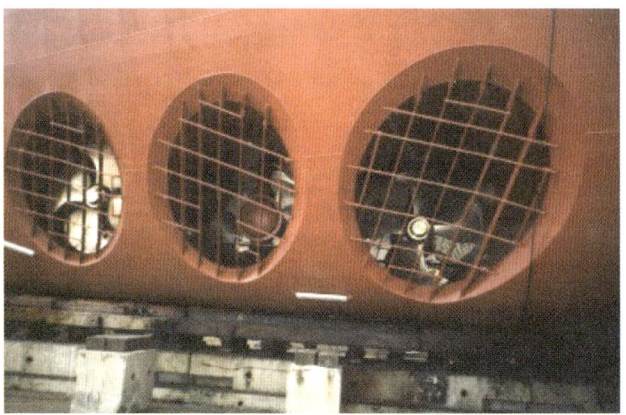

Figure 1.13   Tunnel Thruster (with guards)

Figure 1.12   Seven Oceans - Pipe Lay Vessel

The tunnel thruster is mounted athwartships at the bow or stern inside a cylindrical tunnel. There are two types of tunnel thruster, both of which are most commonly driven by an AC constant speed motor, the L-drive and the Z-drive.

The L-drive type tunnel thruster has the controlling motor sited directly above the tunnel thruster unit. The drive shaft unit is fitted vertically into the propeller shaft.

The Z-drive type tunnel thruster does not have the controlling motor sited directly above the tunnel thruster unit. Therefore a horizontal shaft is required to connect it to the vertical shaft.

Generally the propeller blades will be controllable pitch. Locating shoes on the blades are moved by a sliding yoke mechanism, which alters the pitch as required.

The tunnel thruster shown is the most common unit in use and consists of a controllable pitch propeller (CPP) fitted in a cylindrical tunnel leading from one side of the vessel to the other, perpendicular to the vessels fore and aft centreline. The unit shown is fitted with steel rope guards, which are intended to keep debris clear of the propeller blades. In practice however, the guards can often trap debris within the tunnel thus causing more damage than would otherwise occur.

The unit can be used to manoeuvre the bow or stern to port or starboard by altering the pitch of the propeller blades. The unit can only be used in these two directions, but pitch and power can be varied. If more than one thruster is fitted, as shown in figure 1.13, the thrusters can be controlled individually or operated in unison with a single control.

Tunnel thrusters are most efficient when the vessel is static. As the vessel speed increases, the flow of water through the athwartships tunnel is retarded. This reduction in the direct water flow results in cavitation and a subsequent reduction in thruster efficiency and is one of the main disadvantages of the thruster type. Advantages of tunnel thrusters include the relative simplicity of the system, which results in reduced maintenance, potential breakdowns and spare part requirements.

Azimuth Thrusters

The azimuth thruster has essentially replaced the standard main propeller and rudder manoeuvring system onboard the majority of Offshore Support Vessels. As such the azimuth thruster is extremely common, both as a stern mounted main propulsion system and also as a forward retractable thruster.

The main advantage of such azimuth thusters is the increased manoeuvrability that the fully rotatable system

Figure 1.14 Stern Azimuth Thrusters

provides in static conditions. However, in comparison to the conventional tunnel thruster and conventional main propeller, the azimuth thruster does not have the same level of reliability.

The reasons for this contrasting reliability can be summarised as follows:-

- Increase in draft of the vessel. Although this can be counteracted by the provision of a retractable azimuth thruster, the increase in design requirements and engineering construction of the retractable thruster has its own associated reliability problems.
- Increased exposure to damage.
- Increased requirement for an extra bearing and seal arrangement to allow the rotation of the thruster.

Retractable Azimuth Thrusters

The design and operational function of retractable azimuth thrusters is essentially the same as for stern mounted azimuths. However the retractable azimuth thruster, due to its general positioning at the bow of the Offshore Support Vessel, protrudes below the vessel's hull and therefore requires the added complication of a retraction system. This system allows the unit to be retracted into the hull of the vessel to avoid potential damage when operating in shallow water. As with the standard azimuth thrusters, this type of unit provides increased manoeuvrability for the vessel.

Figure 1.15 Retractable Stern Azimuth Thruster

Propellers and Rudders

Although the majority of Offshore Support Vessels are now fitted with azimuth thrusters as their main propulsion systems, many particularly older vessels still have traditional propeller and rudder combinations. A basic understanding of the systems available and their particular functional capabilities is therefore necessary for any deck officer onboard such vessels.

Figure 1.16 shows the port side main propeller of the Offshore Support Vessel *Seisranger*. The vessel has twin main propellers, three bow tunnel and two stern tunnel thrusters.

Figure 1.16 Seisranger - Main Propeller

The *Seisranger* was previously a seismic survey vessel and has been converted for ROV operations, hence the unusual main propeller manoeuvring system. However, a number of vessels within the industry have similar main propellers and rudders configurations.

As with a standard cargo vessel, the thrust of the single propeller blades can be categorised into two main components. The main component is the fore and aft force that drives the vessel through the water, with an additional and significantly smaller athwartships component, known as transverse thrust.

Main propellers can be provided in a variety of different configurations, with single screw, twin screw, right handed and left handed propellers, controllable and fixed pitch blades all common. It is therefore essential to understand how the particular propeller will react when manoeuvring ahead or astern.

Single Propeller Vessels

A right handed propeller with fixed pitch when providing ahead power will cause a transverse thrust effect with the bow drifting to port.

A right handed propeller with fixed pitch when providing astern power will cause a transverse thrust effect with the bow drifting to starboard.

A left handed propeller with fixed pitch when providing ahead power will cause a transverse thrust effect with the bow drifting to starboard.

A left handed propeller with fixed pitch when providing astern power will cause a transverse thrust effect with the bow drifting to port.

A right handed propeller with controllable pitch when providing ahead power, will cause a transverse thrust effect with the bow drifting to port. It should be noted that as the vessel has a controllable pitch propeller, the unit always rotates clockwise (i.e.- to the right) even when the vessel is going astern, therefore the bias or transverse thrust effect is always to lead the bow to port.

A left handed propeller with controllable pitch when providing ahead power, will cause a transverse thrust effect with the bow drifting to starboard. It should be noted that as the vessel has a controllable pitch propeller, the unit always rotates anti-clockwise (i.e.- to the left) even when the vessel is going astern, therefore the bias or transverse thrust effect is always to lead the bow to starboard.

Twin Propeller Vessels

The added complication of twin propellers adds to the potential possibilities with regards to variations in the system types provided. For example, a vessel can be provided with inward turning propellers. One propeller will therefore be left handed and one right handed. Outward turning propellers may also be provided, again with one propeller left handed and one right handed. Twin propellers turning in opposite directions from each other will effectively cancel out the effects of transverse thrust.

Bridge Design

Bridge design onboard any Offshore Support Vessel, whether involved in diving, ROV, survey, pipe and product deployment or construction work, is directly related to the type of operations and the working environment to which the vessel will be exposed. Specific design features will be required for specific operations and environments; however the Offshore Support Vessels bridge design should take into consideration the following:

Safe Navigation

Irrespective of the specific function of any vessel, the safe navigation and transit between port and location

Figure 1.17 Forward Bridge Console

and from location to location remains one of the primary concerns of any master and his navigating officers. The bridge design of the vessel should therefore ensure that a forward console is provided with the following navigational and transit systems:

- A dedicated manual and auto control station for engine, rudder and thruster control.
- Navigational equipment, including radar, electronic and paper charts, echo sounders and autopilot should be provided, with good all round visibility for navigation and collision avoidance available to the navigating officer.
- External and internal communications systems should be available at the forward console. In the case of Dive Support Vessels, divers may be expected to be in saturation during transit periods, therefore immediate communication between Dive Control and the bridge is essential.
- Monitoring of internal safety systems such as fire alarms and watertight door status should be provided.
- With the introduction of the ISPS Code, the security of the vessel is now a more high level priority and therefore in certain circumstances it may be necessary for camera monitors to be provided on the bridge displaying live feeds from 'critical' areas of the vessel. This may be of particular importance when transiting high risk areas.

Manoeuvring in Port and Offshore

- Generally, Offshore Support Vessels will be set-up and stabilised on DP prior to entering the 500 metre of any installation, therefore manoeuvring at an installation will be conducted in DP mode. However, in port such vessels will generally be manoeuvred in manual control and therefore manual controls must be situated to afford good visibility of the vessel's position. Generally manual and joystick controls will be provided on both bridge wings in addition to the forward and aft consoles.
- Change over controls from manual to auto and from auto to manual must be situated to ensure easy access and avoid confusion. Such controls must be clearly marked.
- Whilst manoeuvring, whether from the main forward or aft consoles or from bridge wing stations, direct access to echo sounder displays, heading indicators, weather indicators and any overside camera displays should be available.
- External and internal communications systems should be available at the forward console and after consoles and at any remote bridge wing stations to allow direct communication between the bridge, engine room and mooring stations.

Dynamic Positioning

It is standard practice for Offshore Support Vessels to position themselves stern to the operations site, dependent of weather and the proximity of obstructions and installations. It is therefore standard practice for Dynamic Positioning operations to be conducted from the after console so that the navigating officers can view the vessel's after deck directly and have a better view of any ongoing operations. However both forward and after DP consoles are the standard on any Offshore Support Vessel.

The bridge design of the vessel should therefore ensure that a forward and after console is provided with the following considerations taken into account:

- Dedicated forward and after control stations should be provided incorporating access to all reference system displays and monitoring systems (such as motion reference units).
- Whilst in DP mode, whether at the forward or aft consoles, direct access to echo sounder displays, heading indicators, weather indicators and any overside camera displays should be available.
- External and internal communications systems should be available at both consoles with immediate and direct access to the engine room, operational control centres (dive, ROV, survey, cranes) and to adjacent installations or other vessels.
- Monitoring of internal safety systems such as fire alarms and watertight door status should be provided.
- Although less risk of security breaches may be expected at offshore locations, the security of the vessel remains a high level priority and therefore it may be necessary for camera monitors to be provided on the bridge displaying live feeds from 'critical' areas of the vessel.

Figure 1.18 Subsea Viking - Forward Bridge Console for Navigation and Transit Control

Monitoring of Diving, ROV, Crane, Survey and other Critical Operations

Whilst on location, an Offshore Support Vessel may be involved in a variety of operations in close proximity to offshore installations where position keeping capabilities and the awareness of the bridge crew is critical to ensure the safety of the vessel's personnel and / or the integrity of adjacent installations and their own vessel.

The bridge design of the vessel should therefore ensure that:

- A direct and unobstructed view of the after deck is provided, including views of any overside launching systems for diving systems, ROV launch systems and cranes.
- Whilst operations are being conducted from the vessel, direct access to echo sounder displays, heading indicators, weather indicators and any overside camera displays should be available.
- External and internal communications systems should be available at both consoles. In the case of diving vessels immediate communication between Dive Control and the bridge is essential.
- Monitoring of internal safety systems such as fire alarms and watertight door status should be provided.
- Although less risk of security breaches may be expected at offshore locations, the security of the vessel remains a high level priority and therefore it may be necessary for camera monitors to be provided on the bridge displaying live feeds from 'critical' areas of the vessel.

Emergency Situations

Emergency situations can, due to their very nature, occur at any time and whilst the vessel may be transiting, manoeuvring or operating alongside an installation. The provision of emergency equipment and systems is therefore generic for most circumstances to allow the navigating officers and master to command and control the vessel during such emergencies.

The bridge design of the vessel should therefore include:

- External and internal communications systems available for direct communications with Dive Control, the engine room, helicopter landing officer, fire parties and emergency teams.
- Monitoring of internal safety systems such as fire alarms and watertight door status should be provided.
- Monitoring and control systems for the vessel's ballast and bilge system should be provided for possible

damage and collision situations where immediate action may be necessary.

Accommodation

Due to the diversity of personnel required to work onboard a modern Offshore Support Vessel, it is now common for not only a marine crew but also ROV, survey, diving and specific project personnel to be onboard the vessel at any given time. Vessels such as the *Subsea Viking* carry a full complement of 70 persons and many vessels of this type will have in excess of 100 personnel. As such the accommodation and services provided onboard are more elaborate and extensive than for the majority of vessels.

Added to the large crew onboard, it is common practice for client representatives to be on the vessel for the entire project duration. As such, they will be able to comment with first hand knowledge on the comfort and facilities onboard and the expectation level has increased correspondingly.

The minimum accommodation requirements for an Offshore Support Vessel now include:

- Dining facilities to allow for the majority of the personnel onboard to dine at the same time, allowing for shift patterns.
- Galley facilities to allow for the storage and preparation of four meals per day whilst on project work.

Figure 1.19 Subsea Viking - Dining Facilities

- Recreation areas for both smoking and non-smoking crew members with satellite television, DVD and video systems and internet access.
- Gymnasium facilities with suitable equipment conducive with offshore conditions.
- Single berth cabins with en suite facilities.
- Dedicated stewards for full accommodation cleaning and laundry services.
- Office facilities for client representatives with full communications and office equipment.
- Meeting and conference suites are also a pre-requisite for project briefings and safety meetings for large numbers of project and marine crew members.

Moonpools

Moonpools on Offshore Support Vessels provide a central, sheltered launch and recovery area for saturation diving bells, ROVs and tooling arrangements. The main purpose of the moonpool being to provide access to subsea, in a location where the sea surface is dampened from the effects of the environmental conditions. In addition, the location of the moonpool on the central athwartships axis of the vessel, where the effects of the vessel rolling has the least effect, provides the system with a more stable platform for operations. If the moonpool is further positioned on the fore and aft axis of the vessel, the effects of the vessels pitching will also have the least effect possible.

Figure 1.20 shows the central ROV launch system on an Offshore Support Vessel. As can be seen from the photograph, the launching and recovery of the ROV through this centrally located moonpool provides protection for the ROV and associated handling system from the ROV hangar until the unit is sub surface. In addition to protecting the system from external environmental effects, the moonpool also ensures that the ROV is not liable to damage from contact with ships equipment or the ships structure, as the ROV can be launched vertically in a central area that has minimal effects due to the pitch and roll movements of the vessel. Similarly, the ROV crews required to operate the handling system for launch and recovery operations, are not exposed directly to the environment, as they can be whilst utilising overside handling systems.

Figure 1.20 ROV Moonpool

In the system shown above, the moonpool has a top hatch arrangement on a trolley / track which can effectively seal off the moonpool when not in use. This area can then be used for storage of the ROV units for maintenance and preparations for the next operation. The ability to seal off the moonpool also ensures that during transits, sea water will not be forced upwards into the working area. Similarly the hatch arrangement, when in place, ensures that the open moonpool area does not pose a safety hazard to personnel working in the area.

A number of different methods are available for sealing off moonpools. Some open work moonpools are sealed by a system of upper and lower hatch covers which interlink with each other and are bolted in place, whilst hydraulically tenting hatch arrangements are also common.

Note: The upper (deck mounted) moonpool hatch cover in figure 1.22 is stored underneath the lower hatch cover. The lower hatch cover is fitted with vertical stanchions for the connection of the two covers.

Figure 1.21 Aft Working Moonpool fitted with Multi-Purpose Handling System (MPHS)

Figure 1.22 Upper and Lower Hatch Covers for Moonpool

A-Frames

A-Frames can be fitted for a variety of purposes such as the deployment of arrays for survey operations.

Figure 1.23 A-Frame Arrangement

Figure 1.24 A-Frame Arrangement

Helidecks

Due to the extended periods that an Offshore Support Vessel may be expected to remain at sea without a port call and the high number of marine crew and project personnel that are onboard during offshore operations, a helideck is now considered a pre-requisite and essential design feature.

Figure 1.25 Offshore Support Vessel Helideck

The design and construction of the helideck will be as per the requirements of the vessel's Classification Society; however the installation must also be approved for operation by the relevant national authority. In the United Kingdom, an approval is required from the Helicopter Certification Agency (HCA) as per the requirements of their guidelines (CAP 437). Similarly most nations have their own requirements and guidelines which must be adhered to, although some moves towards more standardised acceptance criteria are being progressed.

Full details of the requirements for an approved helideck installation in United Kingdom waters can be found in CAP 437, however the following main details have been provided as an overview of the design and requirements for a helideck onboard a generic Offshore Support Vessel.

Location and Structural Strength of the Helideck
- Clear approach and take-off sectors are required for the helicopters, in a position where any structure induced airflow and temperature effects are minimised. Turbulence from hydrocarbon emissions, structures or gas exhausts are considered and may lead to specific limitations on helicopter approach and take-off directions.
- Generally helidecks will be positioned above the highest point of the main structure, however in some cases allowances may be made.
- Minimum obstructions should be adjacent to the helideck and approach and take-off areas.
- The structural strength of the helideck should be designed for the heaviest and largest helicopter that it is anticipated will be utilised.
- The helideck will also be designed to withstand not only standard landings from the heaviest helicopter, but also impact loads from emergency landings.

In general terms the helideck will be provided with:
- A non-slip landing area commensurate with the helicopters for which the helideck is designed.
- The landing area will be marked with the vessel's name, perimeter extremity, obstacle free sector (indicated by a chevron), the helideck D-value, maximum allowable weight and an aiming circle and landing 'H'. Prohibited landing heading sectors should be marked if obstructions dictate such areas.

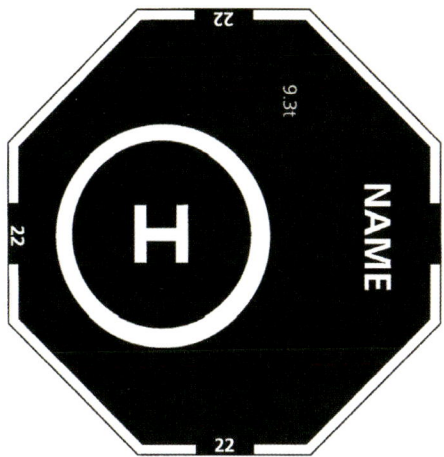

Figure 1.26 UK Helideck markings

- A drainage system for directing fuel spills and rain water from the helideck.
- Tautly stretched rope netting is provided to assist with the helicopter landing and to assist with maintaining the helicopters position once landed.
- Tie down points should be provided for securing the maximum sized helicopter for which the helideck is approved and designed. These tie down points should either be removable or flush fitting to ensure that no obstructions are present on the helideck.
- Safety netting should be provided around the circumference of the helideck for personnel protection.
- There should be a minimum of two entry / exit routes to and from the helideck. These routes should be such that personnel will have an exit route upwind of the landing area, in the event of an incident. The majority of helidecks will have three such routes, two at the aft end of the helideck and a forward escape route.
- A windsock should be provided, indicating the clean area wind directions.
- The safe landing area will be lit with yellow, omni-directional lights, visible above the landing area level. The landing area should be provided with suitable

floodlights if intended for helicopter operations during the hours of darkness. Such lighting shall be installed to ensure that the pilot of any helicopter cannot be dazzled.
- Any fixed obstacles which may affect the landing of a helicopter by imposing a hazard should be provided with daylight hazard markings and omni directional red lights for night time operations.
- Equipment suitable to provide the helicopter pilot with accurate pitch, roll, heave, maximum pitch, maximum roll, maximum heave, temperature, wind speed, wind direction and barometric pressure, in addition to sea state and visibility should be available.
- Communication equipment including a suitable aeronautical radio beacon must be provided to ensure communication between the vessel control station (bridge), helicopter landing officer and helicopter pilot.

Figure 1.27 Helicopter Landing Offshore

Chapter 2

OFFSHORE SUPPORT VESSEL DESIGN — DIVE SUPPORT VESSELS

General Introduction

Chapter 1 introduced the differing vessel types and functions that can be included within the term Offshore Support Vessel and provided some basic details on the generic design features onboard such vessel types. This section aims to expand on this generic design and detail specific equipment and systems that are required onboard a Dive Support Vessel specifically for saturation diving and the associated emergency evacuation of divers under pressure.

for medical evacuation and / or transfer of medical staff and equipment to the vessel in the case of a diver emergency.

- Most Dive Support Vessels will also be fitted with ROV systems. Such systems will be utilised for standalone ROV operations, joint diver and ROV intervention operations and also as an aid to the divers to observe specific hazards.

Figure 2.1   Rockwater 2 – Dive Support Vessel

In generic terms, the Dive Support Vessel will, in addition to the standard Offshore Support Vessel systems and equipment, have the following design features:

- Centrally located saturation dive system including living chambers, diving bell(s), bell handling system(s) and moonpool(s).
- Emergency evacuation system for the dive system including evacuation escape trunking and self propelled hyperbaric lifeboat(s).
- Approved helideck for crew change purposes. However the helideck can provide a suitable platform

Figure 2.2   Diver and Diving bell

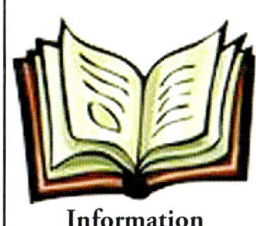

Information

Why do we need to use divers?

As the function capabilities and reliability of remotely operated vehicles has increased, so the need for active on-site human participation in many offshore subsea operations has decreased. However, there are circumstances when a diver is still necessary. This may be due to the location of the activity (where an ROV is physically too large to be deployed into the space), due to the need for manipulation of equipment (that would be too complex for ROV interfaces and manipulators) or where decision making skills are required in real time on-site.

Saturation Dive Systems — Classification Society Design Principles

The Saturation Dive System onboard any vessel is designed within a defined set of principles, relating to the conditions, both environmentally and operationally, within which the vessel and system will be required to function safely. Additionally, the dive system should provide the divers with environmentally controlled living and working spaces with suitable emergency and contingency systems, in the event of an emergency within the dive system or affecting the vessel.

The principles for the design and operation of a dive system are detailed by the relevant Classification Society. For example, Det Norske Veritas (DNV) has specific 'Rules for Certification of Diving Systems'.

DNV Classification	Depth Restrictions	Maximum Operation Time
DSV Class I	The maximum operating depth of the diving system to be equal to or less than 125 metres	The time required from the start of pressurisation until the diver is returned to atmospheric conditions to be equal to or less than 12 hours
DSV Class II	The maximum operating depth of the diving system to be equal to or less than 200 metres	The time required from the start of pressurisation until the diver is returned to atmospheric conditions to be equal to or less than 48 hours
DSV Class III	None, except those imposed by the rule requirements	None, except those imposed by the rule requirements

Figure 2.3 Dive System Class Notations
Det Norske Veritas

Location of the System

A safe and stable location for the dive system is essential. This requirement will take into consideration the need for the system to be operable within set limits of roll, permanent list, pitch and trim. This requirement allows for the safe operation of the system not only in adverse weather conditions but also in situations where the vessel may be damaged and subject to a heel or trim.

Positioning of the dive system amidships on the centreline of the vessel will provide the minimum movement for the comfort of the divers and maintenance of the system and will maximise the horizontal distance between the divers and diving bells and vessel thrusters, propellers and anchors. The central location also provides a protective barrier in the event of a collision or damage to the vessel's main structure. Protection from direct contact with the elements ensures that the environmental conditions within the system can be controlled.

Environmental Conditions

The main function of any dive system is to ensure that the divers are maintained in a safe environment and are provided with comfortable living quarters and working areas. This environment should cater to all basic human needs. Environmental conditions that can affect the launch, recovery and operation of the dive system will mainly be concerned with the sea conditions experienced which will dictate the roll, pitch and heave of the vessel. However, there are additional environmental factors such air temperature and humidity which will also vary considerably with the vessel's location. Dive systems are therefore designed and constructed to function within a range of conditions, as summarised below for DNV classed systems:

Roll, List, Pitch and Trim

For a monohull vessel, the dive system must be capable of being operated safely within the following specified roll, permanent list, pitch and trim ranges.

System	Roll	Permanent List	Pitch	Trim
Chambers and other surface installations	+/- 22.5°	+/- 15°	+/- 10°	+/- 5°
Components in a bell	+/- 45°	+/- 22.5°	—	—

Figure 2.4 Environmental limits for Monohull Vessels
Det Norske Veritas

For a semi-submersible vessel, the dive system must be capable of being operated safely within the following specified roll, list, pitch and trim ranges.

System	Roll	Permanent List	Pitch	Trim
Chambers and other surface installations	—	+/- 15°	—	+/- 15°
Components in a bell	+/- 45°	+/- 22.5°	—	—

Figure 2.5 Environmental limits for Semi-Submersible Vessels
Det Norske Veritas

Temperature, Humidity and Pressure

The living chambers, diving bells and ancillary systems within a diving complex, classed by DNV, will be required to operate safely within specified ranges,

>
> **Reference**
>
> **The Diving Medical Advisory Committee DMAC 26**
> **Saturation Diving Chamber Hygiene**
> **Why are the environmental conditions so important?**
>
> One of the most frequent problems associated with saturation diving is infection. Due to the very restricted and enclosed environment, microbial growth is enhanced and superficial medical complaints such as ear and soft tissue infections are very common. The hygiene levels maintained within the living chamber are therefore of the utmost importance.
>
> Safeguarding against infection within chambers involves control of humidity (which should be maintained at the dry end of the range of comfort), the use of hot water at no less than 60°C for cleaning and meticulous conduct of the onboard procedures to ensure the purity of the fresh water supplies. Samples of potable water should be tested by a laboratory before committing to, and at regular intervals during, saturation.

such as:

- Ambient pressure (1 bar to 1.3 times the pressure corresponding to the maximum operating depth) for all systems;
- Range of ambient temperature (5°C to 55°C) for a living chamber;
- Range of ambient temperature (-10°C to 55°C) for exterior ancillary systems;
- Range of ambient temperature (-2°C to 30°C) for submerged systems such as a diving bell;
- Relative humidity (100%) for living chambers and outer systems;
- Atmosphere contaminated by salt (up to 1 mg salt per 1 m^3 of air) for living chambers and exterior ancillary systems;
- Salinity of ambient water (35 parts per thousand) for submerged systems such as a diving bell;
- Maximum pressure at maximum operating depth for submerged systems such as a diving bell.

Saturation Dive Systems — Operational Design Considerations

The Classification Society rules and recommendations and classification designation are generally concerned with the provision of a suitably safe and technically sound living chamber and diving system in order to maintain the divers at the specified atmospheric pressures. However, there are additional aspects to be considered in the design of a dive system which are of similar importance, such as the practical installation, maintenance and operation of the system.

- The provision of a dive system includes numerous main and ancillary systems including the living chamber, diving bells and transfer locks. All of these systems require valuable space onboard the vessel, therefore the efficient use of the space available is essential in the design and provision of a dive system.
- Large capacities of gas are required to be stored onboard and suitable storage areas must be provided.
- The dive control and saturation control centres should be located in close proximity to the diving bell launch stations and living chambers respectively.
- No known single failure mode should prevent the safe recovery of divers. It is therefore essential that there are back-up systems provided in the case of the failure of the primary launch and recovery systems.
- The hyperbaric evacuation system must be suitably positioned to allow safe and efficient access to the divers from the saturation complex.

Figure 2.6 Toisa Polaris — Class III Dive Support Vessel

Saturation Dive System Components

The configuration and specific equipment which comprises a saturation diving system can vary dependent on a number of factors such as the primary function of the vessel (deep water or shallow water operations), geographical location of the vessel or the type of vessel onto which the system is fitted (mono hull or semi-submersible). However, all saturation dive systems have similar components which perform the same functions regardless of their actual complexity.

The following section therefore details a generic dive system in order to provide a basic overview of the main components within such a system:

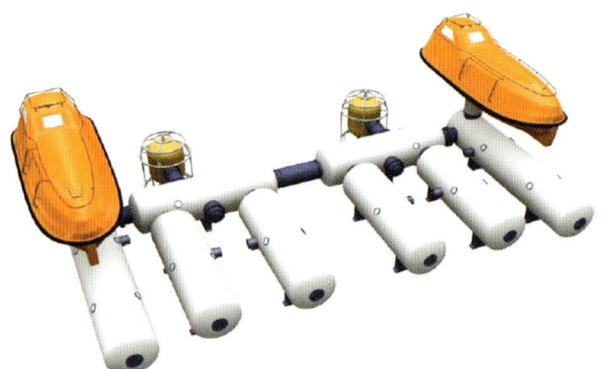

Figure 2.7 Dive System Chamber Complex

(i) Living Chamber Complex

The living chamber complex provides a pressurised environment for the saturation divers' living quarters thus eliminating the need for decompression at the termination of each dive. Each chamber is equipped with the necessary living, sanitary and resting facilities.

(ii) Living Chamber Complex – Life Support Systems and Functions

The life support system comprises the gas supply, breathing gas, pressure regulating and environmental control systems that are required to provide a safe habitat for the divers within the living chamber.

(iii) Transfer Under Pressure (TUP) System

A transfer under pressure (TUP) system is provided to allow the transfer of the divers from the living chamber complex via a transfer lock and mating trunking to the diving bell(s). The TUP system can therefore be considered as the interface between the living and working areas of the saturation system.

(iv) Diving Bell(s) and Diving Bell Life Support Systems and Functions

The diving bell is a submersible diving chamber, which can be pressurised and withstand external pressures, for transfer of divers between the surface living chamber complex and the underwater work site. The systems required for the life support of the divers when within the diving bell and whilst on excursion from the bell are included within this overview. This therefore includes the main bell and diver's excursion umbilicals.

(v) Diving Bell Handling Systems

The diving bell handling system includes all the equipment necessary to connect and disconnect the bell to the chambers and to transport the bell between the surface vessel and the underwater working site. This therefore includes any guide rope systems and cursors.

(vi) Saturation and Dive Control Rooms

There are two main control centres within a saturation dive system; the dive control room and the saturation (or life support) control room. During diving operations, including bell launch and recovery, the dive control room will be manned and all communication between the divers and the surface will be controlled and monitored by the Dive Supervisors. The saturation control room is utilised to monitor the divers whilst in the saturation living chambers and to control all compression / de-compression operations. Life Support Technicians are provided to oversee this area of the operation.

(vii) Emergency Evacuation Systems including Self-Propelled Hyperbaric Lifeboat (SPHL)

The divers living within a chamber complex onboard a Dive Support Vessel will be at various atmospheric pressures, but not at surface atmospheric pressure. They cannot therefore utilise the vessels conventional lifeboats or liferafts in an emergency evacuation. For this reason, dedicated purpose built, self-propelled hyperbaric lifeboats are provided for the divers. A pressurised trunking system is also required in order to provide the divers with a route from the living chamber complex to the deck mounted lifeboat stations.

(viii) Divers Personal Equipment

When on excursion from the diving bell, dependent on sea conditions, the divers will wear a wetsuit or hot water diving suit which is circulated with hot water through the umbilical and divers helmet.

(i) Living Chamber Complex

Figure 2.8 Dive System Chamber Complex

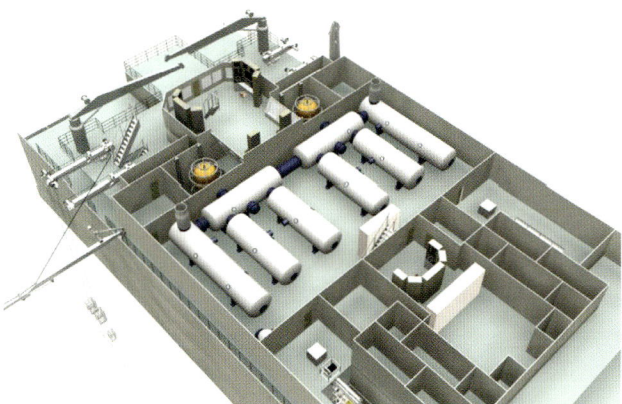

Figure 2.9 Dive System Chamber and Bell Complex

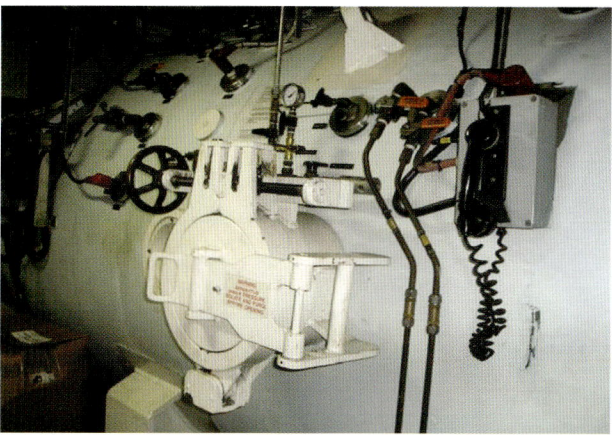

Figure 2.10 Living Chamber – Equipment Lock

There are a number of practical factors that must be considered in the design and layout of the living chamber complex, such as:

- Each living chamber must be suitably accessible for safe operation, maintenance and inspection purposes.
- The living chambers must be provided with suitable trunking for the transfer of the divers to the diving bell(s) and the emergency evacuation systems. Connecting mechanisms between the living chambers and the bell(s) are to be arranged to ensure that they cannot be operated when the trunk is pressurised.
- The living chambers must have sufficient space within for sleeping, eating and sanitary arrangements for a full complement of divers. Space within such living chambers is very limited and minimum space allocations, specified by the Classification Society, must be adhered to. For Class III diving systems, one shower and one toilet are required for each pressure compartment, separate to the central living area. Facilities must be in place to allow the use of one section of the chamber system as a medical treatment area.
- Sanitary systems connected to external systems are to be designed to avoid an unintentional pressure rise in the external system in case of malfunction or rupture of the sanitary systems.
- All living chambers should be fitted with clear windows allowing the life support team to view and visually monitor the divers whilst in saturation.
- The chambers should be fitted with suitable equipment locks for the transfer of provisions and medical equipment while the chamber is pressurised. A safety interlock system must be fitted to the clamping mechanism securing the outer lock door. This interlock door must make it impossible to open the clamp if there is still pressure inside the lock and impossible to pressurise the lock if the clamp is not properly closed.
- Each living compartment is to be equipped with a built-in breathing system (BIBS). Breathing masks corresponding to the maximum number of divers for which the chamber is certified plus one must be provided as part of the system.
- All pipe penetrations in the chambers are to be fitted with external shut-off valves. In addition, non-return valves, flow-fuses or internal shut-off valves are to be mounted.
- The living chambers are to be fitted with a safety valve or a visual or audible overpressure alarm.
- A fully redundant heating system is to be provided enabling temperature regulation to +/- 1°C.
- A system to reduce the humidity in the living chamber is to be provided.
- Internal circulation for gas in the living chambers is to be such that homogeneous gas content is ensured.
- A carbon dioxide removal system is to be provided for each living chamber. This system is to be fully redundant.
- Chamber compartments are to be fitted with indicators visible to the divers, which indicate the internal pressure within the chamber.
- Oxygen analysers are to be provided within the living chambers.
- Direct means of communication is to be provided for each compartment and the life support technicians. A secondary emergency system should also be in place and diving systems intended for the use of helium are to be provided with helium communication un-scramblers.

A number of these considerations require life support systems to be fitted to the chambers. Some of these critical systems are considered in the following section.

Figure 2.11   Living Chamber - Internal

(ii) Living Chamber Complex - Life Support Systems and Functions

The life support systems and functions as detailed within this section are required for the maintenance of life within the living chamber complex. The following are the main life support functions in a generic chamber:

(a)   Breathing Gases
(b)   Gas Storage and Gas Distribution
(c)   Gas Regeneration
(d)   Built-in Breathing System (BIBS)
(e)   Potable and Hot Water Supplies
(f)   Environmental Control Systems (pressure, temperature and humidity)
(g)   Environmental Analysers and Gaugers
(h) CO_2 Absorbers
(i)   Fire-Fighting Systems
(j)   Communication Systems

(a) Breathing Gases

There are a number of different gases that can be used in the various stages of decompression and diving operations. Although the various artificial gas mixtures or pure gases that can be used in saturation diving systems are very different in nature they all share common features:

- All gases used must contain sufficient levels of oxygen to maintain human life;
- All gases used must not contain any gases that are harmful to human life such as carbon monoxide or carbon dioxide;
- All gases used must not become toxic at high pressures such as experienced at deep water depths.

Types of Component Gases

Oxygen

Oxygen must be a component in any breathable gas used for saturation systems as it is essential to sustaining human life. The proportion of oxygen that is used within the gas mixture will determine the depth to which the mixture can be safely used.

- Hypoxic: Gas mixtures which have less oxygen proportions than air (21%) are designed as deep water depth breathable gases. An example of such a gas is Heliox.
- Normoxic: Gas mixtures which have the same oxygen proportion as air (21%) are only capable of being used to relatively shallow water depths. An example of such a gas is Trimix.
- Hyperoxic: Gas mixtures which have more oxygen proportions than air (21%) can be used to shorten decompression times. This is possible as the dissolved nitrogen is drawn out of the body more quickly.

Nitrogen

Nitrogen is the principal component in air. However, the main concern with the use of nitrogen is nitrogen narcosis. As the water depth increases, the level of nitrogen content within a gas mixture that can cause such narcosis decreases. Therefore nitrogen use is limited to shallow water diving operations.

Reference

HSE Diving Information Sheet No. 4   Compression Chambers

What are Safety Interlocks?

Diving systems should be fitted with safety interlocks where necessary, to prevent any unintentional pressurisation or de-pressurisation, or uncontrolled loss of pressure. Particular attention should be paid to chamber / bell mating systems, diver evacuation mating systems, and food and equipment locks. It must be impossible to open the mating clamp between the bell and the chamber while the trunking is under pressure.

What is nitrogen narcosis?

Nitrogen narcosis is attributed to the effects of high nitrogen levels in the nervous system and appears as a type of intoxication. The direct effects of the nitrogen levels and also the resulting irrational behaviour that the intoxication can cause can both lead to death. The effect is also known as the 'Rapture of the Deep'.

Helium

In common with nitrogen, helium is an inert gas and can also, to a lesser extent, cause narcotic problems and decompression sickness. Due to the lesser effects, helium is more suited to deeper water pressures and dives than nitrogen.

Types of Mixed Gases

Air

Natural air is a mixture of 21% oxygen, 78% nitrogen and 1% of trace gases. Due to the high component level of nitrogen in air, it can cause nitrogen narcosis and therefore is only used for shallow depth diving operations.

Nitrox

Nitrox is a mixture of oxygen and air and is used to increase decompression times and reduce the possibility of decompression sickness.

Heliox

Heliox is a mixture of oxygen, nitrogen and helium and is used for deep water diving operations.

(b) Gas Storage and Distribution

The various gases needed for diving operations necessitate the requirement for large quantities of high pressure cylinders (whether fixed bulk storage cylinders or portable quads) to be carried onboard the Dive Support Vessel. As such particular care and attention is required to suitably store the gases with the minimum potential for damage, leakage, fire or explosion.

- All cylinders and quads must be colour coded and marked with the name of the chemical and chemical symbol of the gas contents.
- All diving breathing mixtures should be checked on receipt and re-checked immediately prior to connecting them to a diving gas supply or breathing apparatus charging system.
- All oxygen cylinders and quads must be stored in the open and well clear of any potential fire hazards. This is also applicable to gas mixtures that include oxygen content in excess of 25%.
- In the case of bulk high pressure gas cylinders, an oxygen analyser with visual / audible alarm should be provided. Pressure relief valves from such an enclosed system should be routed clear of the enclosed space.

Figure 2.12   Storage of Gas Cylinders

Gas	Cylinder Body	Cylinder Top
Helium	Brown	Brown
Diving Oxygen	Black	White
Industrial Oxygen	Black	Black
Oxygen-Helium Mixtures	Brown	Brown-White Quarters
Nitrogen	Grey	Black
Air	Grey	Black-White Quarters

Figure 2.13   Marking of Gas Cylinders

What is Decompression sickness?

Decompression sickness (or the bends) is a physiological disorder caused by a rapid decrease in atmospheric pressure when divers return to normal atmospheric pressure too quickly. During ascent to the surface, the decrease in air pressure releases body nitrogen in the form of gas bubbles that block the small veins and arteries cutting off the oxygen supply and causing nausea, dizziness, paralysis, and other neurological symptoms. In severe cases there may be shock, total collapse and death. Gradual decompression allows the nitrogen to be released slowly from the blood and expired from the lungs. Inhalation of pure oxygen aids in clearing nitrogen from the body.

HSE Diving Information Sheet No. 3

Breathing Gas Management

HSE Diving Information Sheet No. 3 provides guidance on the provision, maintenance and use of divers breathing gas in chamber and diving bell systems. Further guidance is provided on the marking and storing of gas cylinders.

Gas distribution is controlled from the saturation control room. From this location the control panel allows gas to be transferred from the gas storage bottles into mixing tanks and enables gas to be distributed to different points within the diving complex. Dedicated gas compressors are provided to facilitate the gas transfer.

(c) Gas Regeneration

Gas regeneration is the process whereby the exhaled gases from divers within the saturation complex can be, following analysis and dilution with fresh gas, re-used for divers breathing gas. Gas recovery and regeneration is possible from all living chambers, bell trunkings and medical and equipment locks.

In order for such gases to be recovered and utilised, they must pass through a gas recovery or reclaim system which comprises a series of filters (scrubbers), a gas reclaim compressor, reclaim bag (tank) and storage tanks. The gas reclaim bag (tank) must be fitted with a suitable monitoring system and relief valve to ensure that there is no risk of over pressurisation.

The series of filters are required in order to remove any contamination, water vapour, bacteria, carbon dioxide, carbon monoxide, light hydrocarbons and methane. The correct and continued operation of the filters is critical to any reclaim system. In order to ensure that they are replaced at appropriate intervals, the reclaim compressor is linked to a run time alarm which will activate in saturation control.

In order to safe guard the divers, a dive system which utilises reclaimed gases will be provided with carbon dioxide and carbon monoxide analysers with audio and visual high level alarms on the divers gas supply. Analysis of nitrogen levels in the system is also essential when using reclaimed gases.

(d) Built-in Breathing System (BIBS)

Oxygen, or any breathing mix, can be supplied to breathing masks inside the chamber. This built in breathing system consists of demand valves for each mask and dump (exhaust) valves for the exhaled gases. The exhaled gas contains a high percentage of oxygen and if exhaled directly into the chamber, oxygen would accumulate rapidly and result in a high risk of fire.

During oxygen breathing, the chamber atmosphere is analysed and flushed regularly to maintain safe levels.

Figure 2.14 Built-in Breathing System (BIBS) face mask

It is essential that a breathable gas mixture is maintained on-line to the chamber and to the BIBS at all times. In each compartment of the chamber there must be one BIBS connection and mask for each intended occupant plus one spare.

(e) Potable and Hot Water Supplies

Infection is a major and frequent concern for divers in saturation. The enclosed environment, temperature, humidity, hyperoxia and helium environment contribute to enhanced microbial growth. The control of such growths is essential in order to restrict the possibility of a wide variety of ear, respiratory tract, skin and stomach infections.

Potable or hot water supplies can be the source of many such bacteria. It is therefore essential that the water supplies, particularly for potable water, are monitored closely. Daily and regular checks should therefore be performed to analyse the water supply including conductivity, pH, turbidity, chlorine, mineral content (iron, copper and zinc) and bacterial coliform and thermostable coliform counts.

Hot water should be provided at no less than 60°C to provide maximum protection against possible bacterial growths.

(f) Environmental Control (Pressure, Temperature and Humidity)

There must be suitable means of controlling the internal environment within the living chamber complex including the pressure, temperature and humidity. A secondary system should also be available for the provision of heating and humidity control.

With regards to pressure, the chamber depth must be maintained accurately at all times. The pressure difference between the gas storage bank and the chambers should be such that, in the event of a sudden reduction in chamber pressure, the loss may be maintained by pressurisation from the storage banks. In addition, there must always be a minimum of two supplies of pressurisation gas on-line to the saturation control panel.

The chamber temperature should be maintained between 28°C and 34°C, but ultimately at a temperature that is comfortable for the divers.

The control of the humidity within the chamber complex should be maintained at the dry end of the range of comfort. Humidity greater than 65% in a chamber environment can provide ideal conditions for infections of the skin and external ear.

Each lock within the chamber complex will be equipped with temperature and humidity sensors.

(g) Environmental Analysers and Gauges

Environmental analysers and gauges are available for a variety of purposes from a number of specialist companies. These analysers and gauges include oxygen content, temperature, carbon dioxide content, depth and humidity.

(h) CO_2 Absorber

Carbon dioxide absorbers or scrubbers are designed to remove carbon dioxide from the chamber atmosphere. Such units can consist of a fan that routes the chamber air through a canister filled with absorbent such as Sodasorb.

(i) Fire-Fighting Systems

Each chamber should be fitted with an internal fire deluge system. Such deluge systems will comprise a fire deluge water storage reservoir, internal water deluge nozzles, internal activation cord and saturation control panel activation controls.

(j) Communication Systems

Communication between the life support technicians and the divers is a considerable problem. Normal speech is distorted due to the large amounts of helium used in the breathing gas mixtures used during compression. The effect of pressure and breathing gas on the diver's voice patterns becomes more pronounced as the pressure to which they are exposed increases. This situation is obviously an important factor in the general welfare of the divers and for emergency situations where clear communication is essential. As such, diving saturation systems are provided with speech unscramblers for divers underwater and for divers within the living chamber system.

The system operates using a frequency modulation method to modify the divers' speech to a comprehensible level and can be utilised for divers at more than one depth if more than one modulation unit is provided. An additional system allows divers at different depths to communicate utilising a cross talk system so that a diver at 300 metres can speak to another diver at 100 metres depth.

(iii) Transfer Under Pressure (TUP) System

The transfer under pressure (TUP) system allows the divers to be transported from the shipboard living chambers to the subsea or seabed worksite whilst remaining at the same depth pressure. The living chamber will therefore be maintained at the working depth pressure and the divers will be transferred via a trunking system to the diving bell and then to the seabed, without the need for any decompression.

In a modern saturation dive system, there may be three living chambers, all of which may be being maintained at different pressures. A separate transfer lock or chamber is therefore provided. When one set of divers are required to deploy to the worksite, the transfer lock or chamber and diving bell are pressurised to the same level as their respective living chamber. Once the system pressure is balanced, the divers can be transferred via the chamber and bell mating hatches.

The access to the diving bell will be via the bell trunking from the transfer lock or chamber. The bell mating clamp incorporates a safety interlock and two double pressure locking hatches to prevent the clamp being opened until the chamber to bell trunk is vented.

Typically, there may be a safety hydraulic pressure switch on the hydraulic system which prevents the bell clamp hydraulics being operated until the bell trunking

Figure 2.15 Bell Mating Trunking and Clamp

has been vented.

A bottom / top transfer mating system, as shown in figure 2.15, is the most common arrangement on monohull vessels, but a side mating system is also in use on some Dive Support Vessels, particularly semi-submersible units.

(iv) Diving Bell(s) and Diving Bell Life Support Systems and Functions

The diving bell is essentially the means of deploying the divers from the living chambers to the worksite and as such is the interface between the vessel based and the subsea systems. The diving bell should be considered as an extension of the living chambers and as such, the diver's environment must be similarly controlled whilst in the diving bell. The positioning, construction, life support systems and dive equipment are all therefore critical in maintaining the safety and relative comfort of the divers during launching, recovery and dive operations.

Positioning of Diving Bells

The positioning and configuration of the diving bell or bells and subsequently the dive system, is heavily dependent on the ship motion characteristics and the operational sea states that the vessel will be expected to operate in. There are a number of different configurations and, as with most vessels; there will be differences in all systems. However, the main arrangements for a two bell diving system can be summarised as:

Two Diving Bells on the Centreline

With two diving bells installed on the centreline of the vessel, one will generally be located amidships thus being very stable. The second bell will be sited forward or aft of midships and therefore will be more prone to movement. In such instances, the bell positioned midships will be the preferred option, but both will be relatively stable. However, as both diving bells are on the centreline of the vessel, there will be practical restrictions. As the diver's umbilical will effectively be deployed from the centreline of the vessel, the diver's excursions limits will be reduced due to the distance required to reach the extremity of the vessel. This would further be reduced if the worksite, such as an offshore platform, has a considerable overhang, with the diver having to contend with the distance from the centreline of the vessel to the side shell, the distance between the vessel and the platform and the subsequent overhang.

Two Diving Bells off the Centreline of the Vessel (on same side)

This configuration would reduce the diver excursion required to deploy the diver to the side of the vessel and would therefore put the diver closer to the dive site. In certain circumstances, the arrangement would also eliminate the need to cross haul the bell to the side of the vessel with the ships crane, when alongside offshore platforms. However, due to the bells both being located off the centreline of the vessel, the effects of the ships motion would be increased and therefore the launching and recovery of the bells would be conducted in a less stable environment.

Two Diving Bells Side by Side

An arrangement with a port and starboard diving bell would allow the vessel to operate port side or starboard side to the platform or structure, with the diving bell closest to the work site being used. By having the option to use either bell, there will be more flexibility for vessel operations and in either case, the diving bell and therefore the divers will be closer to the worksite on deployment. However, the positioning of the diving bells off the centreline would increase the effects of the vessels motion. A further disadvantage to the fore and aft bell configuration is the limitation on bell usage when alongside a site. It may only be possible to launch one bell during such periods, with the bell furthest away from the site not being utilised.

Irrespective of the configuration used for the dive system and diving bells, the common purpose of any such arrangement will be to:

- Locate the bells as close to the centreline of the vessel as possible to lower the effects of roll, heave and pitch from the vessel.
- Position the divers as close to the worksite as possible by maximising the length of the diver's umbilical in order to achieve the greatest working distance. However, it should be noted that, for safety reasons, the length of the divers umbilical is set at 30 metres.

- Ensure that the bells can be launched and recovered in specified sea states and environmental conditions for the intended area of operation for the vessel.

Figure 2.16 Saturation Diving Bell

Construction of Diving Bells

Diving bells are designed to provide a safe environment for the divers in both operational and emergency situations and a means of transferring them from the surface to the worksite. The construction of diving bells therefore encompasses both safety and practical considerations for the deployment and subsequent recovery of the divers. Diving bells are commonly constructed to accommodate three divers (two working and one bellman or standby diver).

- Diving bells are made of high strength, fine grain steel.
- The insulation of the bell shall be of non-water absorbent cellular type material in the form of non absorbent blocks in resin with a glass reinforced thermostatic (GRP) coating. The metal surfaces of the side hatch, bottom hatch, port hole surrounds, side mating flange and bottom hatch surround are typically made of stainless steel.
- All hull penetrations shall be fitted with double hull valves.
- Anodes are fitted externally and internally for galvanic protection of the diving bell.

- The bell exterior and sub systems (such as gas cylinders) are protected by a bumper ring.
- The bell design should avoid protrusions and sharp edges which may lead to injuries to the divers or, in extreme situations, snagging on underwater obstructions.
- The bell design must allow for the divers to exit and re-enter the bell freely. This may take the form of a bell stand-off below the bottom opening which shall support the divers during exit and entry to the bell. This shall also prevent the obstruction of the bottom entrance should the bell rest on the seabed.
- A method must be available whereby the bellman or standby diver can recover an unconscious diver in to the bell. This will normally be a self locking pulley which can be attached to the divers harness. A partial flooding system should also be fitted to allow the bell to be part filled in order to assist re-entry into the bell of an injured diver.
- The bell design should result in a low centre of gravity.
- A main lift pad eye is provided for connection to the bell hoist wire. A secondary lifting point attachment should be available on the bell in the event that the primary lifting point attachment is damaged.
- An umbilical attachment lug is provided for connection to the main bell umbilical.

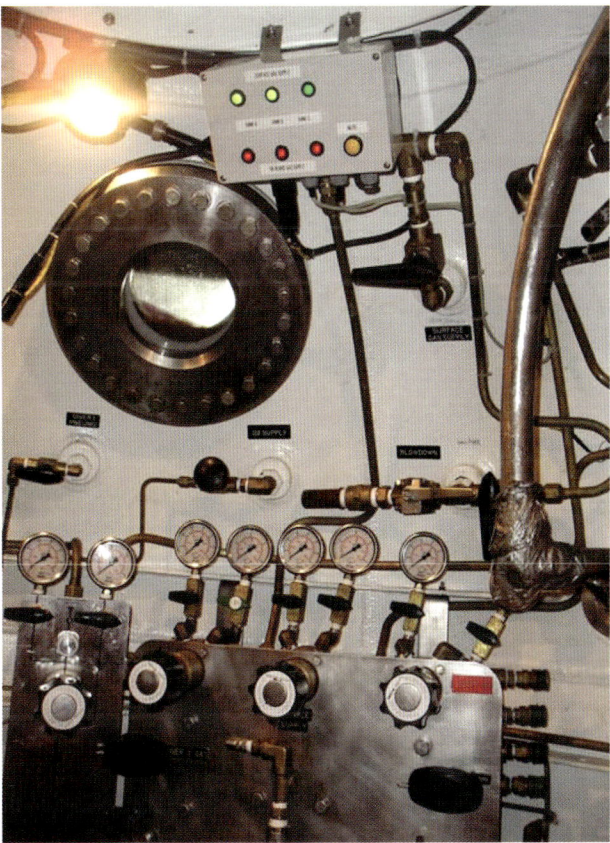

Figure 2.17 Internal Diving Bell Control Panels

- The bell must be provided with emergency gas and oxygen supplies, emergency battery power, emergency heating systems, carbon dioxide scrubber, emergency lighting, built-in breathing system and emergency transponder.
- The bell will either be designed as a positively buoyant or a negatively buoyant chamber. Positively buoyant bells will allow ballast to be attached to the bell, which can subsequently be released during an emergency to allow the positively buoyant bell to return to the surface.

Diving Bell Main Umbilical

The main umbilical is the link between the diving bell and the surface and supplies the primary breathing gas, electrical power, diver's hot water, pneumatics, reclaim gas and communication systems to the diving bell and is therefore a vital link to the surface. In addition, the main umbilical must be strong enough to recover the bell if the main wire breaks.

Figure 2.19 Diving Bell Main Umbilical Winch

Figure 2.18 Diving Bell Main Umbilical Cross-Section

Colour coding is generally utilised for the main components of the main umbilical with gas (yellow), pneumatics (green or blue), hot water (red), reclaim gas (black), communications (red) and video (orange) being the standard colour codes. The outer sheathing is normally orange in colour and can be polyurethane sheath or braided.

Divers Gas Supply

As stated previously, the diver's gas is supplied to the bell via the main umbilical and thereafter via individual diver's umbilical to each diver.

The following details the main requirements of the diving bell gas supply and the diver's excursion umbilical supplies:
- For the occupants of the bell, if the main surface gas supply fails, a shuttle valve will automatically switch over to the onboard breathing mixture. This switch is linked to an audible and visual alarm which will warn the bellman.
- The onboard breathing mixture must be able to support one diver outside of the bell for at least 30 minutes at a breathing rate of circa 42 litres per minute at the maximum depth of the diving operation.
- There must be a primary gas supply for the bellman, which can be from onboard bottles or from the surface, sufficient to allow him to exit the bell and recover an injured diver. This supply must be independent of the primary gas supply to the diver(s) in the water. The bellman must also have a secondary supply; however this may be common with the working diver's primary supply, provided it is protected if the working diver line fails.
- An oxygen analyser must be in place with an audio / visual low / high content alarm on the gas supply to the divers.
- A BIBS mask must be provided for each occupant of the diving bell. This should be capable of providing breathing gas either from the surface or from the onboard cylinders. Both primary and secondary gas supplies must be separate from the supply to both the bell and diver in the water. The supplies must be arranged such that if one line fails then this does not interfere with the supply to other lines.

- For the diver's excursion umbilical, each diver's gas supply must be arranged so that a failure does not affect the other diver's gas supply.

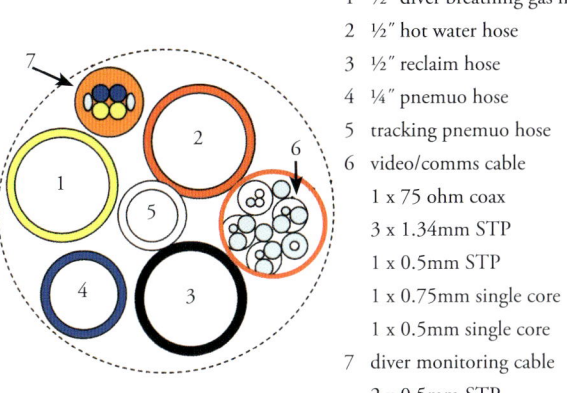

1	½″ diver breathing gas hose
2	½″ hot water hose
3	½″ reclaim hose
4	¼″ pnemuo hose
5	tracking pnemuo hose
6	video/comms cable
	1 x 75 ohm coax
	3 x 1.34mm STP
	1 x 0.5mm STP
	1 x 0.75mm single core
	1 x 0.5mm single core
7	diver monitoring cable
	2 x 0.5mm STP

Figure 2.20   Diver's Umbilical

Bell Ballast Release System

It is a generally accepted best practice (and in some countries a legal requirement), that saturation diving bells are provided with a primary, secondary and alternative means of recovery. The primary means is the main lift wire, with the secondary system being the guide wire system. The use of a bell ballast release system (bell weights) may be an acceptable alternative means of bell recovery, in instances where the main and/or secondary recovery systems are unavailable.

There are many reservations within the diving industry with regards to the use of such ballast release systems with a number of concerns related to the use of such systems.

- The potential for an accidental release of the ballast, resulting in the inadvertent positive buoyancy of the diving bell can be considered a major hazard to the diving bells integrity and therefore the divers. The release mechanism will therefore consist of a primary and secondary activation control, both of which will independently require to be activated before the ballast will release.
- The hazards associated with the release of the buoyancy and therefore the ascent of the diving bell can be considerable. Hazards immediately above a stricken bell, such as severed guide wires and main umbilical, can impede the bells ascent and the proximity of surface obstructions such as the Dive Support Vessel, must also be considered.
- Additional equipment and systems associated with the ballast release system must also be added to the weight of the bell, to allow the buoyancy mechanism to operate successfully. This additional equipment and systems will require even more buoyancy to be fitted. This can result in the bell being difficult to deploy in normal circumstances, due to the increased buoyancy fitted.
- The additional weight of the buoyancy and associated equipment and systems increases the weight of the diving bell as a unit and will therefore lead to an increase in the loads placed on the diving bell handling systems.

These numerous and quite considerable considerations have resulted in many diving contractors not utilising such systems. Alternative means in areas such as the North Sea are available for safe recovery of the divers, with the use of ROV units and also by using the second bell on two bell systems.

AODC 56 and AODC 61

AODC 56 and AODC 61 provide general guidance on the hazards associated with ballast release systems and also guidance on the initial and periodic examination and test of such systems.

Reference

Heating and Emergency Heating Systems

One of the main environmental considerations for divers within a deployed diving bell and for divers while on excursion from the diving bell is temperature. For certain areas of operation, such as the North Sea, the provision of a suitable heating system for the divers is essential to support life, with a primary and secondary means being provided. A primary hot water supply to both the diving bell and the divers on excursion is provided via the main umbilical and diver's umbilical and a means of monitoring the temperature.

Failure of the primary heating system requires the secondary or emergency system, which should prevent excessive heat loss and hypothermia for a period of 24 hours at the diving systems maximum operating depth, therefore providing life support while the recovery of the bell is attempted.

Generally, this secondary or emergency heating system can be achieved by heating the bell environment, the divers directly by heated suits, or by passive thermal insulation by means of passive or active heating systems.

Passive Heating Systems

Passive emergency heating systems are utilised to reduce heat loss and conserve heat, rather than to provide heat. Although a number of different types of such systems are available, there are generic features that are common to each type:

- A CO_2 scrubber with diver mask both retains the heat from the diver's exhaled breath and also from the chemical reaction in the CO_2 scrubber when subject to the diver's exhaled carbon dioxide.
- Thermal insulation suits and bags can reduce heat loss from the divers.

Passive heating systems have a number of advantages over the more complex active heating components, including the following:

- Passive heating systems are very simple and uncomplicated in design, therefore there is very little onboard maintenance required and a very low possibility of failure.
- Once donned, passive systems operate without any active participation by the divers and as such will continue to function even in the case of the diver becoming unconscious or medically incapacitated.

Active Heating Systems

Active heating systems are external sources added to the diving bell in order to provide generated heat. This generated heat is then utilised either directly or indirectly to provide heat to the divers.

As these active systems are fitted externally to the bell, the advantages differ quite substantially to those for the passive diver heating systems:

- No space is taken up internally within the diving bell for the active heating systems.
- The divers' movements are not restricted by active systems, whereas passive systems require the divers to be connected to the face masks.

However, one main similarity does exist between passive and active systems, in that both operate without any active participation by the divers and as such will continue to function even in the case of an unconscious or medically incapacitated diver.

AODC Guidance Note No. 026
Diver Emergency Heating
AODC Guidance Note No; 26 provides basic background and guidance information on the types of passive and active heating systems available for divers.

Reference

Analysers and Sensors

Analysis of the environment within the diving bell can be considered even more critical than for the living chambers; as a failure with divers on the seabed may be considerably more difficult to control and rectify due to the remoteness of the divers.

The monitoring of the diving bell environment to ensure the diver's safety is therefore essential and as such the diving bell should be fitted with similar analysers and scrubbing units as the living chambers. These indicators, which should be visible to the divers, will display the external and internal pressures, pressure of gas storages on the bell and oxygen levels. A carbon dioxide scrubber unit to provide primary removal of the gas from the bell atmosphere will also be provided.

Emergency Equipment

In addition to the emergency breathing (BIBS) and emergency (passive and active) heating systems that have been covered in the proceeding sections, the diving bell must also be provided with suitable systems for the survival of the divers during a lost bell scenario, to assist in the location of the bell, if lost, and for the subsequent connection of a lost bell to third party life support functions.

Survival kits for divers can include both hardware and insulation components. The hardware component can consist of a thermal regenerator unit, passive carbon dioxide scrubber, food, water and sanitary bags. The thermal regenerator unit is connected to the survival bag. The divers exhaled gases pass through the carbon dioxide scrubber, producing heat via a chemical reaction with a sodalime reservoir, with the heat being passed to the regenerator unit and then to the survival bag.

The insulation component can consist of an undersuit specially designed to provide thermal protection and a survival bag, which can be considered as more of a sleeping bag arrangement, having arms and being fitted with a waterproof outer layer. Such survival bags are fitted with integrated harnesses in order to provide support for a sleeping or unconscious diver.

The diving bell is fitted with an emergency locating beacon designed to assist personnel on the surface in establishing and maintaining contact with the submerged diving bell if the umbilical to the surface is severed. A strobe light with a minimum operating duration of 24 hours must also be fitted to the bell to assist in location in an emergency.

For the provision of life support, a manifold must be supplied for the connection of basic supplies including hot water, breathing gas and communications.

A self-contained emergency through-water communication system is also essential.

Emergency Power

Arrangements must be in place to ensure that sufficient power is available for the safe completion of a dive and recovery of the bell if the vessel power fails.

Any equipment identified as necessary to satisfy this requirement must be able to continue operating in the event of a loss of the vessels primary power source.

(v) Diving Bell Handling Systems

The function of the diving bell handling system is to provide the means of transporting the diving bell from the stowed position to the deployment position and to the worksite. Subsequently the handling system is utilised to return the diving bell from the worksite to the stowed position. On two bell systems, each diving bell will be provided with a dedicated handling system.

Bell handling systems differ greatly from one saturation dive system to the next, however in order to simplify the function of the bell handling system and the process required to deploy the diving bell, a generic system can be divided into the following main components:
- Moonpool Cursor
- Transverse Trolley Arrangement
- Guide Wire Weight and Winch Arrangement
- Main Bell Lift Wire and Winch Arrangement

Moonpool Cursor

When diving operations are not being conducted, the diving bell is rigidly clamped to the moonpool cursor. The moonpool cursor can be described as a frame support which maintains the bell in a stable position during transfer from the mating position to the deployment position in the moonpool. The clamping arrangement restrains the bell from the effects of the pitch, heave and roll of the support vessel during such transfers and is attached to the trolley system, as shown in figure 2.21.

Figure 2.21 Saturation Diving Bell

Transverse Trolley Arrangement

For transporting the bell and moonpool cursor from the mating position with the living chamber complex to the deployment position in the moonpool, there is a trolley arrangement. Permanent support beams are installed in the bell hangar deck head, running from the mating position to the moonpool. The overhead trolley is mounted on rails to these support beams and powered hydraulically, with the cursor attached to the trolley mounted winch via, typically, a four part wire system.

The operation of the system to transfer the diving bell from the mating position to deployment will therefore consist of the following stages:
- The cursor is clamped in position at all times during the stowage of the bell and is therefore in place during the transfer of the divers from the living chamber complex into the bell.
- With the divers safely transferred into the bell and bell mating clamp disengaged, the overhead trolley is utilised to transfer the bell to the centre of the moonpool.
- With the bell positioned above the moonpool, the cursor and trolley mounted winch is utilised to lower the bell into the deployment position through the moonpool.
- At the lower extremity of the moonpool, the submerged weight of the bell is transferred to the main bell winch system.
- The bell can then be released from the cursor clamp arrangement and deployment continues utilising the bell main lift wire.

Guide Wire Weight and Winch Arrangement

A guide wire system will consist of a guide weight; guide wire rigged in two parts and a guide wire winch and sheave arrangement. Such an arrangement is installed for a number of reasons. During deployment and recovery of the diving bell, the system restricts excessive lateral and rotational movement of the bell in the water, with the guide weight acting as a stabilising force.

The guide weight is lowered to the seabed underneath the diving bell and acts as both a storage basket for tools and equipment, but also ensures that a safe access point is available for the divers exiting from the bell.

In the event of a failure of the main bell lift wire, the bell can be recovered by utilising the guide wire arrangement.

Typically, this arrangement will only lift the bell in water.

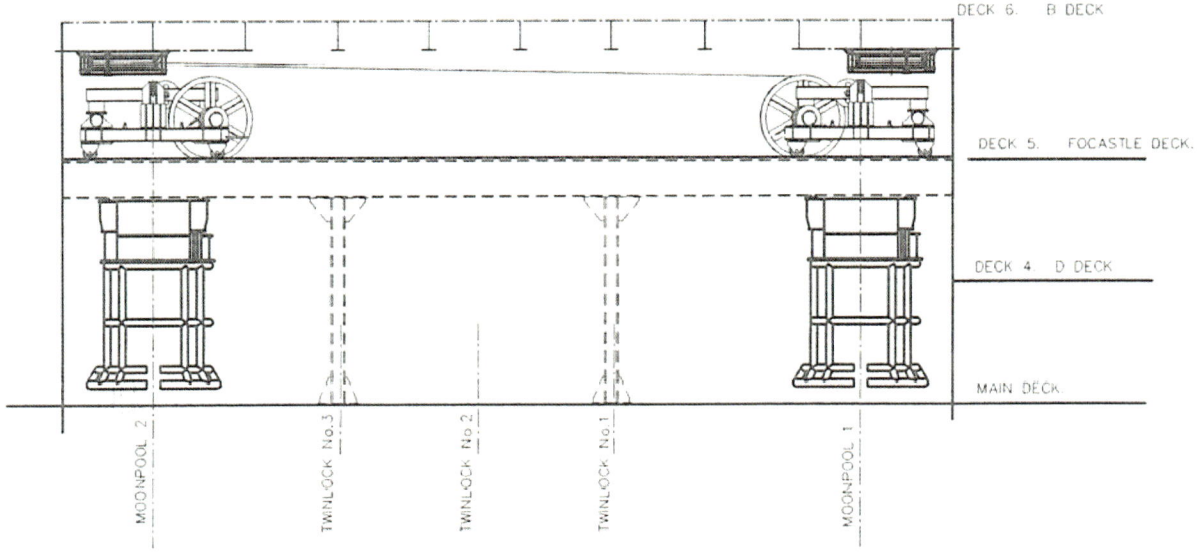

Figure 2.22 Schematic of Diving Bell, Moonpool Cursor and Overhead Trolley Arrangement

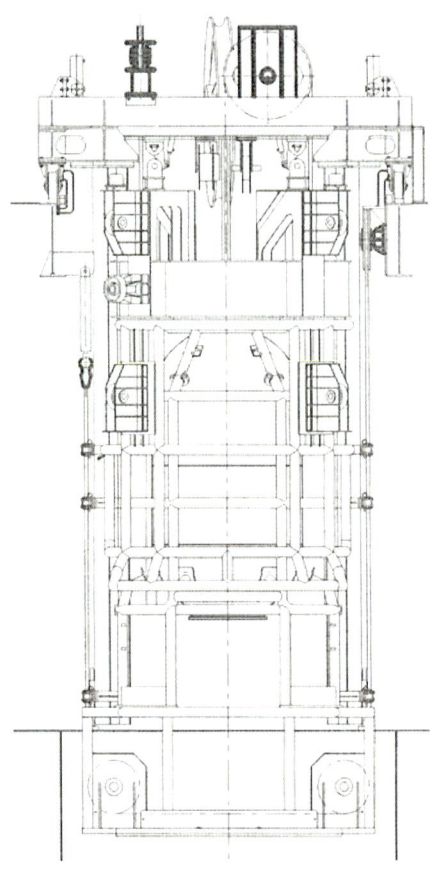

Figure 2.23
Schematic of Diving Bell, Moonpool Cursor and Overhead Trolley Arrangement

Figure 2.24   Guide Wire Weight

Main Bell Lift Wire and Winch Arrangement

The launch and retrieval of the diving bell presents significant hazards to the divers, particularly during heavy weather, and are major factors in the configuration and operation of the handling system. With regards to the main bell lift wire, it is essential that the winch arrangement is designed with the capability to pass the diving bell through the air-surface interface at sufficient speed to avoid excessive wave action impacting the bell and should have sufficient lifting capabilities to allow fast retrieval of the diving bell. Suitable and sufficient controls and brakes which permit precision control for bell mating must be provided.

It is generally recommended that the main lift wire should be non-rotating and that galvanised wire should be used.

(vi) Saturation and Dive Control Rooms

Saturation Control Room

The control and monitoring of the divers and their living environment whilst they are in the saturation living chamber complex is conducted by the Life Support Technicians and controlled via the saturation control room. Additional controls and monitoring is also provided for the evacuation trunking to the hyperbaric lifeboat.

The saturation control room must therefore contain all the systems required for the maintenance of all life support functions for the divers within the complex as summarised below.

- **Gas Regulation** – Panels are provided for the regulation, mixing and transfer of oxygen and gas mixtures to the chamber system.
- **Oxygen and Carbon Dioxide Analysis** – A gas analysis panel is provided with inlets from each of the gas supplies, allowing each supply to be subject to oxygen and carbon dioxide analysis. Audio and visual alarms are provided for the oxygen and carbon dioxide levels in the chambers. An oxygen monitor will be provided in the saturation control room atmosphere in case of leaks.
- **Humidity Control** – Control of the chamber temperature and humidity is provided with data on the internal environmental conditions in the chamber system being provided by temperature and humidity sensors.
- **Depth Monitoring** – The pressure depth for each compartment may be provided with high and low alarms. Such systems are also fitted with indications of the rate of change within any compartment so that the change of pressure the divers are subject to can be controlled to a high degree of accuracy. Pressure depth alarms are fitted for high and low levels and rates of change.
- **Miscellaneous Monitoring** - Reclaim gas bag (tank)high level alarm, SPHL battery fan and general shipboard emergency alarms and signals are provided.
- **Communication Systems** – Primary and secondary two way communication between the saturation control room and critical positions onboard, such as each internal chamber, the bridge, each exterior equipment lock, hyperbaric lifeboat launch stations, compressor rooms and gas storage areas, is essential.
- **Emergency Systems** – BIBS are provided in the saturation control room to enable the Life Support Technicians to remain at their control stations during emergency situations, whereby the atmosphere of the control room may be compromised. Emergency lighting must also be provided to allow continued operations in the case of a blackout.
- **Hyperbaric Lifeboat Control Panel** – Controls and monitoring equipment is provided for the transfer of the divers from the living chamber complex to the hyperbaric lifeboat during an evacuation. These controls will include trunk pressure and exhausts, lifeboat oxygen content and depth gauges and communication systems between the saturation control room and the trunking and the lifeboat chamber. Visual monitors will be available for viewing the interior of the lifeboat chamber.

Dive Control Room

All operations associated with the deployment of the divers from the surface living chamber to the worksite and their subsequent return, are controlled and monitored from the dive control room. Dive control therefore includes the following control and monitoring systems:

- **Gas Regulation** – Panels are provided for the control of gas supplies, exhaust gases and gas reclaim systems to and from the diving bells. A separate panel or separate dive control room will be provided for each diving bell within a saturation system.
- **Oxygen and Carbon Dioxide Analysis** – There must be a means of monitoring the diving bell atmosphere for oxygen and carbon dioxide levels. An oxygen analyser with audio / visual high / low alarms must be provided to warn the occupants of any rise or fall in the oxygen content outside pre-set parameters due to gas leakage in the area.
- **Depth Monitoring** – Monitoring equipment is provided to display the pressure depth for the diving bell, depth of each individual diver and the supply

pressures of each main and back-up breathing gas supply.

- **Visual Monitoring** - Monitors for each diver's helmet mounted cameras plus slave monitors from critical locations, such as the ROV station and survey data displays should be available in dive control.
- **Alarm Systems** – The dynamic positioning alert / alarm (green, yellow, red) and ship's emergency alarms must be duplicated in dive control. A mute function for the ship's alarms must be provided to allow critical operations and communications between dive control and the divers in an emergency situation. Dive system alarms for high and low hot water temperature, diver's gas high and low oxygen levels, diver's gas high reclaim carbon dioxide alarm, dive control room oxygen atmosphere alarm and miscellaneous other alarms are provided.
- **Communication Systems** – Primary and secondary means of communication between the bridge and dive control must be provided. The primary communication system must be hard wired and dedicated for direct communications and should be fitted with a back-up emergency power source, such as batteries. Through water communication between dive control and the bell, when the bell is deployed must be provided. A sound powered telephone should be provided as the secondary means of communication between the divers in the bell and dive control. Communications must be available between dive control and other critical positions onboard, such as the bridge, crane operators and ROV control room. All communications between the divers and dive control must be recorded.
- **Launching and Recovery Control Stations** - Ideally, dive control will be arranged so that direct visual contact is provided of the launch and recovery operations of the bell and for mating operations between the bell and the living chamber complex. However, monitors should be provided for each bell handling system.
- **Emergency Systems** – Emergency breathing apparatus with communication facilities must be available for the dive control supervisor and for the bell hoist winch operator so that recovery and life support functions can continue in a hazardous atmosphere.

(vii) Emergency Evacuation Systems including Self-Propelled Hyperbaric Lifeboat (SPHL)

General Introduction

The whole purpose and function of a saturation diving system is to maintain divers at a predetermined pressure depth, in order to allow prolonged and repeated

Figure 2.25   Saturation Control Room

Figure 2.26   Dive Control Room

working excursions at that depth. However, one major disadvantage that is incurred with maintaining the divers at such pressures is the problem of how to evacuate them from the surface living chambers and subsequently from the vessel. Due to the problems associated with rapid changes in pressure environments on humans, it is not practical to decompress the divers to the surface pressure within a short period of time, as would be required during an emergency. It is therefore necessary to provide the divers with a suitable escape system which can be maintained at a similar pressure depth and which can be used solely by the divers in saturation.

These systems take the form of hyperbaric lifeboats, which externally appear very similar in design to standard enclosed lifeboats. Connected to the living chamber system by way of an escape trunking arrangement and clamp connection, the hyperbaric lifeboat includes an internal pressure (living) vessel for the evacuated divers and the associated life support systems and functions with the trunking arrangement being maintained at the same pressure as the main living chamber to ensure that a safe and timely evacuation is possible in an emergency situation.

Self-Propelled Hyperbaric Lifeboats

A generic hyperbaric lifeboat can be sub-divided into two main areas:

- External Structure and Marine Equipment
- Internal Chamber and Life Support Functions

Figure 2.27 Self-Propelled Hyperbaric Lifeboat

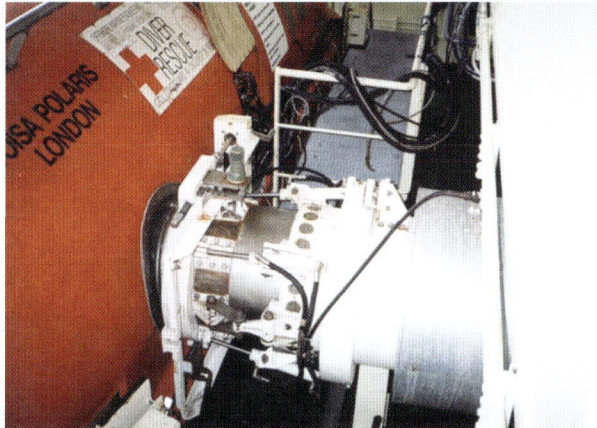

Figure 2.28 Hyperbaric Lifeboat Mating

External Structure and Marine Equipment

The external lifeboat structure is designed in a similar manner to a standard enclosed lifeboat and must therefore meet SOLAS and IMO requirements and standard design criteria, in addition to complications associated with the living chamber and the divers.

- A strength test must be conducted to ensure that the lifeboat has sufficient strength to enable it to be lowered into the water when in the fully loaded condition and to ensure that it can be launched and subsequently towed.

- A stability test should demonstrate that the lifeboat has positive stability in the water when fully loaded and also in the specific flooded conditions.

- An impact test, when in a simulated fully loaded condition, will be conducted. With the lifeboat in a free hanging position, it should be pulled laterally to such a position that when released it will strike a rigid vertical surface at a velocity of 3.5 m/s. It should then be released to impact against the rigid vertical surface. No damage should be sustained that would affect the lifeboat's efficient functioning.

- A drop test should be conducted with the lifeboat suspended above the water line so that the distance from the lowest point of the lifeboat to the water is 3 metres. The lifeboat should then be released so that it falls freely into the water. No damage should be sustained that would render the lifeboat unserviceable.

- A self-righting test, in a simulated fully loaded and the light condition should be performed. The lifeboat is incrementally turned to angles of heel up to and including 180º and should, on being released, always turn to the upright position.

- A flooded capsizing test is performed with all exits in the open position. A self-righting test is conducted with the lifeboat rotated about the fore and aft axis incrementally up to and including 180º. On release, the lifeboat should attain a position that provides an above-water escape for the occupants.

- The hyperbaric lifeboat requires an access connection for the divers from the main dive system living chambers to the internal lifeboat chamber. Once the hyperbaric lifeboat has been disconnected from the evacuation trunking, the connection access remains extruding from the lifeboat structure. This access connection will be used to evacuate the divers once the lifeboat has reached a suitable rescue centre with saturation chambers. As such, the lifeboat has extra weight on the side of the trunking, which is counterbalanced within the lifeboat structure itself.

- Fire protection of the lifeboat consists of a water spray system and an air support system is provided for the marine crew and Life Support Technicians sufficient for 10 minutes supply. The lifeboat structure should be made of fire resistant material and suitable protection for the breathing gas bottles and piping systems should be provided.

- Dedicated hyperbaric lifeboats should be coloured orange and be provided with retroreflective material to assist in their location during periods of darkness. The external canopy should be marked in accordance with IMO and IMCA (D027 – Marking of Hyperbaric Lifeboats) with a minimum of three identical signs, one of which should be clearly visible from air rescue units (figure 2.29).

- All external connections such as those for emergency gas, hot and cold water and communications should be clearly marked. Lifting points, maximum gross weight, parent vessel and port of registry and emergency contact details should also be displayed.

- Warning signs should be displayed to warn any potential rescuers who do not have specialised diving knowledge and experience, not to touch any valves or other controls, try to remove the occupants from

Figure 2.29 Hyperbaric Lifeboat – External Marking

Figure 2.30 Hyperbaric Lifeboat — Lifting Beam

the chamber or open any hatches, connect any gas, air, water or other supplies and not attempt to give food, drinks or medical supplies to the occupants.

- A tow line may be fitted for use by a rescue vessel or to assist in towing the lifeboat clear of the ship's side. If fitted with a towing line, the line should be secured in such a manner that the line cannot be inadvertently deployed, but is easy to deploy when required.

- For navigational and search and rescue purposes, the lifeboat is required to be equipped with a compass, a radar reflector, automatic flashing light and emergency radio beacon. Although not a requirement, the provision of a handheld global positioning system (GPS), navigational charts for the area of operation and contact and position details for the nearest rescue centre and safe ports should be provided as standard.

- Due to the fact that the divers within the hyperbaric lifeboat chamber may require a prolonged period of decompression, it may not be possible to remove them from the chamber on arrival at a safe haven, following an evacuation. It must therefore be possible to recover the lifeboat from the sea, for subsequent transfer to a hyperbaric rescue centre so that the life support functions can be provided externally to the system. In the case of standard forward and aft lifting point systems, this requirement can be fulfilled by the provision of a lifting beam arrangement which can either be stored onboard an adjacent installation during diving operations, or ashore at designated contingency locations, available for immediate transfer in the case of an emergency situation.

Internal Chamber and Life Support Functions

The hyperbaric lifeboat is required in order to provide a protective platform for the pressure chamber and therefore the divers until rescue can be facilitated. The hyperbaric lifeboat pressure chamber within this protective platform has a single function; to sustain life by maintaining thermal balance. As such it is accurate to describe the lifeboat pressure chamber as a smaller version of the dive system living chamber complex, but with the added complications of being fitted to a much smaller and therefore less stable vessel. The reduction in size from a dive system chamber to a lifeboat chamber also poses problems with regards to available space, provision of facilities, heating and medical assistance in such a confined space. This limited space also reduces the availability of life support system essentials such as breathing gases, heating and food and water supplies.

The following systems and equipment may be expected to be provided in a generic hyperbaric lifeboat chamber and life support console:

- Mixed gas quantities sufficient to compensate for the use of the food and medical locks and to allow for minor leakages.

- Controlled oxygen supplies for at least 24 hours for the maximum amount of divers the lifeboat is designed to accommodate, based on a consumption rate of 0.5 litres per minute.

- There should be one built-in breathing system (BIBS) connection and mask for each occupant of the hyperbaric lifeboat plus one spare. The exhausts from the BIBS should be piped outside of the hyperbaric lifeboat.

- Monitoring of oxygen levels and partial pressures of carbon dioxide should be available and carbon dioxide scrubbers and emergency scrubbing systems fitted.

- To ensure that the divers are maintained in thermal balance, heating (or cooling) systems must be provided for the maximum number of divers available. If powered by mechanical means (lifeboat engine) then this should be capable of running for at least 24 hours. Back-up equipment including thermal clothing, thermal protective aids and emergency survival packs will be provided as standard.

- An emergency generator can be fitted to supply secondary heating and cooling and power for the carbon dioxide scrubbers.

- A fire-extinguishing system should be fitted in the lifeboat chamber suitable for operation in all pressure depths up to and including the maximum pressure depth that the diving system is designed and approved for.

- A medical lock and a food lock should be provided. Safety interlocks should be fitted to the clamping mechanisms to ensure that it is impossible to open the clamp if there is still pressure inside the lock and impossible to pressurise the lock if the clamp is not properly closed.

- Food and water supplies should be adequate for at least 72 hours, but may vary dependent on the environmental conditions of the vessel's operating area and the geographical location and consequently the potential for a timely rescue.
- A hyperbaric toilet with a safety mechanism to ensure against a skin seal. The build up, storage and removal of faeces, urine and vomit are essential to minimising the contamination and risk of disease to the divers in such close confinement. Removal or bagging of waste is therefore essential.
- Communications between the divers in the lifeboat chamber and the lifeboat support crew in the lifeboat control station will be provided with an unscrambler system and back-up system.

Evacuation, Launch and Recovery of the Hyperbaric Lifeboat

The evacuation, launch, manoeuvring and subsequent recovery of the hyperbaric lifeboat must be controlled by personnel other than the divers within the chamber and therefore involves a great deal of reliance on the support and third party personnel.

The escape trunk is maintained at the working pressure depth during routine operations and therefore allows a rapid evacuation to the hyperbaric lifeboat.

The divers exit the living chamber complex, with the last diver ensuring that the access is closed behind him.

With the divers in the lifeboat chamber, the lifeboat access is closed and the escape trunk vented. The venting of the trunk will be possible from the hyperbaric lifeboat launch control position and saturation control. The spool piece, which connects the trunk to the

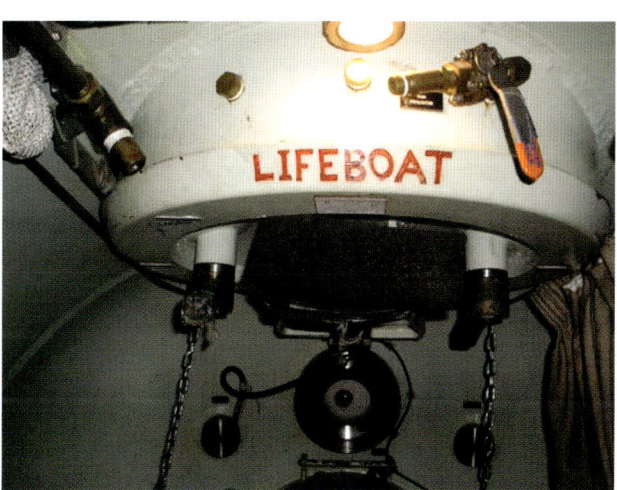

Figure 2.31
Hyperbaric Evacuation Trunk Access from Living Chamber

hyperbaric lifeboat, is retracted and the launching procedure can commence. A safety interlock system is fitted to the clamping mechanism and spool piece between the hyperbaric lifeboat flange and the flange on the connection to the chamber system. This interlock ensures that it is impossible to open the clamp if there is still pressure inside the trunk and impossible to pressurise the trunk if the clamp is not properly closed.

The launching requirements and capabilities for a hyperbaric lifeboat are the same as for a conventional lifeboat and will therefore comply with the SOLAS requirements. However, special care and attention is required in order to ensure that no damage is possible to the side mating flange, which will be required when the lifeboat has been recovered, for the eventual recovery of the divers from the chamber.

The recovery of the hyperbaric lifeboat is unlikely to be performed by the mother vessel and therefore the crane or davit recovering the lifeboat is unlikely to be designed for the purpose. Due to the potential for damage, the hyperbaric lifeboat should be recovered by utilising a dedicated, purpose built spreader beam or frame which will ensure that the stress during lifting will be concentrated on the specially strengthened areas of the lifeboat hull. Limits for lifting will be expressed as maximum significant wave heights; however the limiting sea states will vary from cranes of different capacities, stiffness and hoisting speeds.

Due to the constraints associated with recovering the lifeboat onboard a third party rescue vessel; it is more likely that the lifeboat be escorted to a safe haven with the third party vessel providing shelter in the form of a lee to aid the comfort of the divers during the transit. However, stern recovery of a hyperbaric lifeboat may be possible by new specially designed EERVs.

Transit to a safe haven should form part of the company's standard emergency procedures and therefore the destination should be pre-determined prior to the launch of the lifeboat, with a number of specially designed rescue centres being available in the world. These centres have the capabilities to provide the lifeboat with emergency supplies, life support functions and decompression for the divers.

(vii) Diver's Personal Equipment

Numerous manufacturers within the offshore industry provide a variety of diving suits, diving helmets and associated accessories. However, the main components of the diver's personal equipment when on excursion from the diving bell is the diving helmet and the diving suit. Hot water diving suits are used in cold water environments where hot water is circulated to the suit via the diver's umbilical, with the flow of the hot water controlled by the diver.

The diving helmet design and construction is to provide functional use for the diver (whilst also providing suitable protection for the diver. The use of reinforced fibreglass is common, providing impact resistance whilst remaining lightweight).

The helmet will be fitted with the main and auxiliary gas supplies via a demand regulator, lights, communications and video feeds.

Figure 2.32   Diver's personal equipment — helmet

Chapter 3

OFFSHORE SUPPORT VESSEL DESIGN — ROV SUPPORT VESSELS

General Introduction

ROV operations are often usually simpler and are safer than using human intervention by means of saturation or air divers and ROVs have the advantage of being capable of being deployed for longer periods than their human counterparts. They can also be used in situations where it would be hazardous or inappropriate to send a diver, due to obstructions or adverse weather conditions. However, there are disadvantages in using ROVs instead of divers as the human presence is often lost, making visual surveys and evaluations more difficult.

Many operations can be performed by using a combination of divers and ROVs and therefore most Dive Support Vessels are fitted with ROVs as standard, but often the ROV is seen as an alternative method of conducting the sub surface operation and as such, smaller ROV Support Vessels are sometimes used for ROV only operations.

ROV operations are varied and can include the following:

- Diver Observation – The ROV can be used to observe a diver or dive team to ensure diver safety and to provide assistance for diver operations where particular tools or manipulators may be required.
- Installation Inspection, Cleaning and Debris Removal – The ROV can perform visual inspections and monitoring of offshore installation corrosion, fouling, cracks and structural integrity. Removal and cleaning of fouled debris can also be performed once identified.
- Pipeline Inspection – The ROV can be used to inspect the terminations, manifolds and lengths of underwater pipelines to check for leaks and the pipeline structural condition and integrity.
- Seabed Surveys – Prior to the installation of subsea pipelines, cables and structures, visual surveys of the seabed are conducted by ROV to check for obstructions and potential hazards.
- Drilling Support – An ROV is generally maintained onboard offshore drilling rigs to assist with general drilling operations and more particularly to assist with potential problems. The ROV can be used for visual inspections, monitoring and for repair operations.
- Subsea Installation Construction Support – The ROV can be utilised to assist with the installation and maintenance of manifolds, subsea structures and platforms.
- Telecommunications Support – Similar to pipeline inspections and seabed surveys, the ROV can be utilised to inspect proposed telecommunication cable positions, assist ploughing and cable installation operations.
- Location and Recovery – The most publicly aware use of the ROV is in the location, identification and investigation of ship or aviation wrecks. Offshore Support Vessels are often chartered for such operations, particularly in instances where the water depths involved may be too excessive for diver intervention.
- Pipe Lay Support — The ROV can be used to provide visual support when deploying pipelines, cables or flexibles.

Figure 3.1 Kommandor Subsea – ROV Support Vessel

The ROV Support Vessel will have the following generic design features:

- Centrally located ROV launch and deployment systems. As with dive systems, the centre line of the vessel amidships, provides the most stable area for the recovery and deployment of the ROVs. This can take the form of moonpool systems, generally located in ROV hangars, and also external or internal side launched systems. Internal side launched systems will be located inside the ROV hangar with side doors provided in the ships hull to facilitate recovery and deployment. External systems will tend to be of the A-Frame type. These external systems are also generally temporarily installed and can be removed to provide deck space for other, non-ROV, operations.
- A-Frames, although fitted on a variety of vessel types, will generally only be permanently fitted on purpose built ROV / Survey vessels. The stern area of vessels which have survey vessel capabilities including A-Frames may be provided with a stern roller with detachable stern rails. These can be removed and the A-Frame / stern roller arrangement utilised for towed array systems.

- Survey operations require a high level of data collection and review of the raw information obtained. Survey vessels tend therefore to have large office areas dedicated to the processing of such data.
- A crane for light construction operations.

Classification of ROVs

Class I – Observation ROVs

These vehicles are relatively small units fitted with camera, lights and sonar only. They are primarily intended for pure observation, although they may be able to utilise one additional sensor (such as cathodic protection equipment) as well as an additional video camera.

Class II – Observation ROVs with Payload Option

These vehicles are fitted with two simultaneously viewable cameras and sonar as standard and are capable of handling several additional sensors. They may also have a basic manipulative capability.

Class III – Workclass Vehicles

These are vehicles large enough to carry additional sensors and / or manipulators and commonly have a multiplexing capability that allows additional sensors and tools to operate without being 'hardwired' through the umbilical system. These vehicles are generally larger and more powerful than Class I or II vehicles and wider capability, depth and power variations are possible.

Class IV – Towed and Bottom Crawling Vehicles

Towed vehicles are pulled through the water by a surface craft or winch. Although they do not have propulsive power they may be capable of limited manoeuvrability. Bottom crawling vehicles use a wheel or track system to move across the sea floor, although some may be able to 'swim' limited distances. These vehicles are typically large and heavy and are often designed for one specific task, such as cable burial.

Class V – Prototype or Development Vehicles

Vehicles in this class include those still being developed and those regarded as prototypes. Special purpose vehicles that do not fit into one of the other classes are also assigned to class V. This class includes autonomous underwater vehicles (AUVs).

Types of ROVs

For ROV Support Vessel operations, workclass and observation class units are the standard. Many specific

Figure 3.2 Autonomous Underwater Vehicle

types are in use throughout the industry, but three of the most common units, the Hercules and Centurion HD workclass vehicles and the Seaeye Tiger observation class vehicle are described here to give a general overview of the type of unit that may be found on such vessels.

Workclass ROVs

The main function of a workclass ROV is to facilitate work often associated with diver operations. In situations where the use of divers is precluded due to operational constraints, proximity of hazards and environmental conditions (currents), the workclass ROV can be substituted for a diver and used for a variety of tasks utilising the unit's manipulator arms. As such, workclass ROVs can be used to manipulate, lift and control valves and other subsea equipment.

Hercules (Workclass ROV)

A summary of the design features and capabilities of a standard Hercules system is provided below:

- High manoeuvrability of the ROV is possible, due to the provision of a total of eight thruster propulsion units. Four vertically mounted and four axial mounted thrusters provide the capability for the unit to be manoeuvred horizontally, vertically and laterally.
- The ROV Control System utilises fibre optic technology to provide data and video links from the unit to the operator station. This feature allows the interfacing of tooling equipment and survey sensors to the system.
- The robust but relatively lightweight framing allows good protection of the ROV systems, sensors and indicators. Current Hercules systems are rated up to 3000 metres water depth.
- The ROV is fitted with gyro (auto heading) and pressure (auto depth) control systems, allowing for a high degree of position accuracy. Position fixing of the unit is provided by removable / interchangeable transducer systems.

- Installation of a maximum of six video cameras is available for inspection work and for guidance for manipulator operations.
- A maximum of seven manipulator arms are available for a wide variety of tools and operations.

An optional tether management system can provide additional excursion capability from the main launch position, to allow the ROV to be deployed from a vessel in a safe area, but where close access is required to a platform / offshore structure.

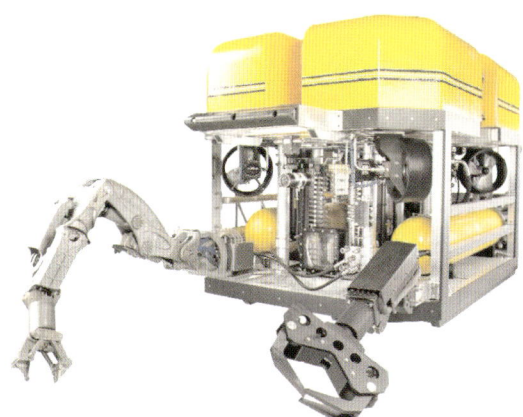

Figure 3.4 Centurion HD Workclass ROV

Seaeye Tiger (Observation Class)

A summary of the design features and capabilities of a standard Seaeye Tiger observation class ROV system are provided below:

- Excellent manoeuvrability is provided, even in strong currents, by four vectored and one vertical thruster units, allowing the unit to be manoeuvred horizontally, vertically and laterally.
- Operations are possible in water depths of up to 1,000 metres.
- The frame is made from polypropylene and is therefore maintenance free and non-corroding. The open frame design allows increased efficiency of the thruster units.
- Auto heading and auto depth functions.

Figure 3.3 Hercules Workclass ROV

Centurion HD (Workclass ROV)

A summary of the design features and capabilities of a standard Centurion system is provided below:

- High manoeuvrability of the ROV is possible, due to the provision of a total of six thruster propulsion units. Four horizontally mounted and two vertically mounted thrusters provide the capability for the unit to be manoeuvred horizontally, vertically and laterally.
- The robust but relatively lightweight framing allows good protection of the ROV systems, sensors and indicators. Current Centurion systems are rated up to 1500 metres water depth.
- The ROV is fitted with gyro (auto heading) and pressure (auto depth) control systems, allowing for a high degree of position accuracy.
- Installation of a maximum of four video cameras is available for inspection work and for guidance for manipulator operations.
- A maximum of seven manipulator arms are available for a wide variety of tools and operations.

An optional tether management system can provide additional excursion capability from the main launch position, to allow the ROV to be deployed from a vessel in a safe area, but where close access is required to a platform / offshore structure.

Figure 3.5 Seaeye Tiger Obsevration Class ROV

- The capability to fit and utilise a single function manipulator such as a cutter unit or for the use of a four function gauntlet manipulator.
- Two tether management systems are available providing the option to increase the standard excursion range from 140 metres to 240 metres from the ROV garage. A 'free swimming option' is also available with no TMS.
- For observation operations, the unit is fitted, as standard, with colour and low light black and white cameras and scanning sonar. The camera units are mounted on a tilt unit. Dual video channels and digital stills capability are provided.

ROV Components and Sub Systems

The aim of any ROV is to be able to deploy to a pre-determined position, conduct the operation required and return to the vessel in a safe manner. The combination of manoeuvring systems (thrusters), auto heading and auto depth controls, position transducers, manipulator arms and video link capabilities make this aim possible. The ROV can therefore be considered as a sum of the various components and sub systems that are required to make up the complete system.

Although every type and make of ROV unit will vary, the following main basic components and sub systems will be considered:

- ROV Frame
- Buoyancy and Ballast Control
- Propulsion Systems (Thrusters)
- Position Keeping Systems
- Cameras, Video Functions and Lighting
- Manipulators, Interface Tools and Skids

ROV Frame

The main purpose of the frame is to provide protection for the more sensitive components, such as the cameras and hydraulic and electric systems and additionally provides a base for the mounting of all system components.

A combination of a requirement to be robust enough to restrict damage to the unit components and light enough to ensure manoeuvrability, aluminium is generally used for the frame. The open nature of the frame will provide the required protection when operating in close proximity to large offshore structures whilst providing an open framework which ensures that system maintenance is not impeded and thruster efficiency remains high.

Figure 3.6 Workclass ROV frame

Buoyancy and Ballast Control

When designing an ROV, the use of lightweight components to keep the overall vehicle weight within practical limits is a major consideration. The combined weight of the ROV will include the weight of the frame, system components, payload and buoyancy modules.

A positively buoyant ROV can be operated anywhere in the water column, and will return to the surface if a power failure occurs or if the tether is lost. This is of particular significance when operating with only one ROV. If there is a failure in the tether, then it will be possible to recover the unit. In addition, a positively buoyant unit can manoeuvre on the seabed without the need to thrust vertically upwards, which will stir up the seabed surface and sediment, thus reducing visibility for the ROV.

To provide this positive buoyancy, modules are fitted to the frame. Such buoyancy modules are fitted to the upper parts of the frame, with the heavier components (such as electric motors) being fitted low to maintain a stable unit with a low centre of gravity.

Figure 3.7 Buoyancy Modules

Neutral and negative buoyancy are also possible, with neutral buoyancy being the most efficient, as the minimum thrust will be required to maintain the vehicle at the requisite depth. Negatively buoyant (heavy) ROVs are also operated as standard and can provide less risk of loss. On vessels that have multiple ROV systems and crane assistance, a heavy ROV lost to the seabed could easily be recovered.

There are two types of ballast in general use on standard offshore ROVs; fixed ballast or variable ballast.

Fixed Ballast

Fixed ballast (positive fixed buoyancy) of a vehicle is achieved by pressure resistant buoyancy chambers or syntactic foam to bring the vehicle to the desired specific gravity. Most vehicles use a syntactic foam block near the top of the vehicle to gain positive buoyancy with a fixed buoyancy displacement.

To avoid significant loss of buoyancy in the case of an impact, it is conventional to use multiple compartments in the frame.

Fixed payload on the vehicle is usually in the form of several lead blocks. This lead may be exchanged for equipment without adjusting the vehicle's foam package.

Variable Ballast

Variable ballast is less common than fixed ballast, but may be used for operations where the vehicle must be neutrally buoyant such as for prolonged seabed work including pipeline and cable burial and repairs.

This type of ballast is provided by open or closed tanks and results in a variable displacement of buoyancy. Open (soft) or closed (hard) tanks may be provided whereby high pressure air is blown into the tanks to displace water, or vented to allow water to enter. This system allows the ROV to be negatively buoyant and therefore ideal for seabed operations, whereby the ROV can manoeuvre without thrusting downward, which would be the case with fixed positive buoyancy.

Propulsion Systems (Thrusters)

The propulsion units or thrusters in an ROV provide the horizontal, vertical and lateral movement.

There are two main types in use; electro-hydraulic or electric. Generally, the weight and relatively lower efficiency of an electro-hydraulic system effectively eliminates this system from consideration in small vehicles. In larger vehicles, however, it has the advantage of simplicity, versatility, reliability and low electrical noise. Electric motor systems weigh less and therefore have an advantage for smaller ROV units.

The maximum static thrust is important for determining the manoeuvrability of the unit, however the system efficiency, and in particular, the location of the thrusters being a major design consideration. The location of the thrusters is important as the velocity of the water surrounding the thrusters, essentially the inlet velocity, affects the output of the thrusters.

Position Keeping Systems

The majority of ROV operations require the ROV to maintain a fixed position with regards to depth, heading

Figure 3.8 Centurion ROV - Thruster Units

and positional location. In addition, the control of pitch and roll of the unit will ensure a stable platform for the intended work role.

In order to provide such station keeping capabilities, ROVs are fitted with the following systems:

- Auto Heading and Auto Depth Control
- Pitch and Roll Control
- Acoustic and Tracking Sensors

Cameras, Video Functions and Lighting

Essential to both workclass and observation class ROV functions is the ability for the ROV operator to be provided with a visual image of the working area and the immediate environment in real time. Apart from the operator's requirement to view the immediate site area in order to control and position the ROV unit, a visual record is the main purpose of many operations, such as pipeline and seabed surveys where the client requirement is to identify and chart any damage or obstructions. Visual identification and positioning is paramount. For workclass vehicles engaged in manipulation work, the ability to view the extremities of the manipulators must also be considered.

To fulfil these requirements, the ROV may be fitted with video and still cameras, mounted on pan and tilt assemblies to allow vertical (tilt) and horizontal (pan) adjustment of the viewing angle.

For the navigation and positioning of the ROV, the use of video can cause problems in low light levels, similar to human vision, whereby depth perception is non-existent. Use of two cameras to provide more than one perspective and therefore a sense of depth is standard, although most units will have more. By utilising an ROV with additional camera units, there can be an added redundancy, provided by overlapping the fields of view of the cameras. Therefore if one unit fails, another can be adjusted to view the intended area. Cameras fitted to the rear of the unit provide a view of the umbilical or tether and can therefore alert the operator to any potential or actual snags or fouling.

Still cameras provide higher quality resolution single images to these video camera systems. However, good quality, but lower resolution, still photographic images can be obtained from the video images. One of the main advantages of using the video imaging is the certainty that what the operator is viewing is what is being recorded.

The option of using colour or low light cameras will be dependent on the operation being carried out, with low light video cameras providing long distance viewing, whereas colour video cameras provide good contrast, but will require extra illumination. Extra illumination can lead to high back scatter of light which can distort the images received.

Both the video and still camera functions of the ROV require light to illuminate the subject matter, whether it is the seabed, a pipeline or subsea structure. High intensity lights are therefore fitted as standard to augment and compliment the video and camera units and will also be fitted on pan and tilt assemblies to provide full coverage of the working area and manipulators.

Manipulators, Interface Tools and Skids

Manipulators and tools are fitted to the ROV in order to provide the base unit with the capability to perform manipulation of subsea objects.

Fitted to the forward end of the unit, the manipulators and interface tools can replace the need for human intervention in environments that may be inhospitable to divers due to the depth, current or potentially hazardous obstructions. Remote manipulation is therefore possible allowing the ROV pilots on the surface vessel to use the mechanical arms and tools in real time, without placing the divers in a potentially hazardous environment. In contrast to the type of robotic arm that may be found in shore based industries, assigned to a single, repeated function, the subsea manipulation arm is more likeable to the mechanical equivalent of the diver's arms and hands and can operate in an ever changing environment and can adapt to different work functions by providing multiple degrees of freedom.

There are numerous manipulator types designed for a variety of underwater tasks from simple, one stage operations, such as holding a lift line to much more complex tasks such as manipulation of electrical or hydraulic connectors and valves. The type used will therefore depend on the task to be completed and will depend on the lift capacity required and number and type of functions needed. The manipulator arm can be further modified and adapted by use of special tooling interfaces for specific operations; the manipulator arm being used as the necessary means to control the tooling device, which will be connected to the manipulator arm extremity.

Non-destructive testing (NDT) sensors are used to ascertain the structural integrity of subsea structures and can include cathodic potential probes and thickness measurement devices. For cleaning operations offshore, rotating wire or nylon brush attachments and water jetting tools can be used.

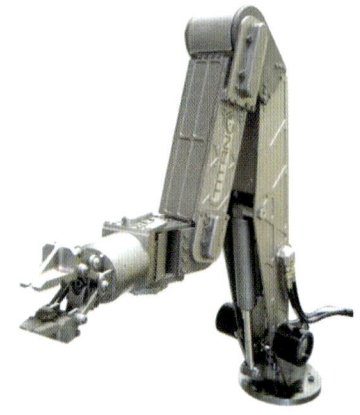

Figure 3.10 Titan 4 Manipulator

Work tools can include simple hand tools operated by the manipulator to multi function tools that require additional hydraulic or electrical power.

In some instances, where the tools required for a particular offshore ROV operation are extensive and task specific, a full tooling spread, rather than manipulator arm attachments may be the preferred option. In these cases, tooling skids can be provided which consist of a bottom mounted attachment, with custom designed manipulators and tools integrated into the skid base. As shown in Figures 3.11 and 3.12, these skids are connected to the base of the ROV and operations continue with the ROV and skid operated as a single unit. Tooling skids are available for a variety of ROV operations, such as gasket removals, drill cuttings removal, hydrocarbon sampling and wellhead cleaning.

ROV Management and Control Systems
Main Umbilical and Tether Management System

Figure 3.9   ROV Unit with Manipulators

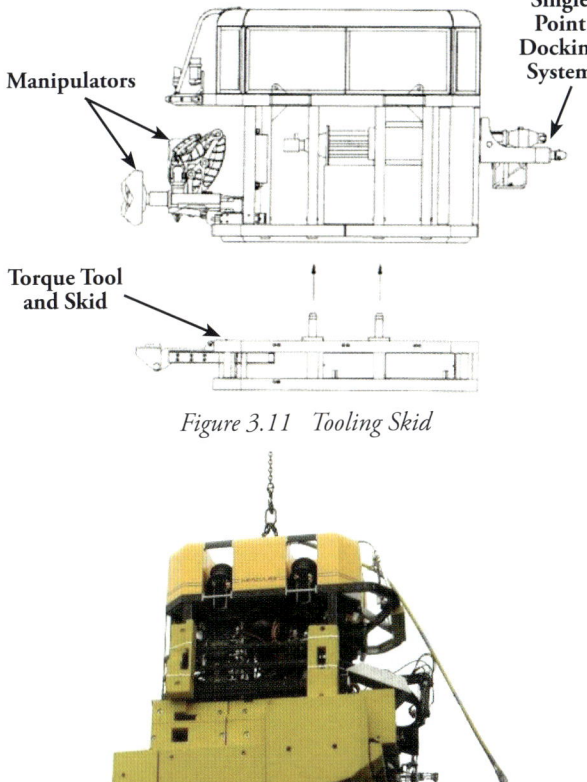

Figure 3.11 Tooling Skid

Figure 3.13 ROV Deployment and Main Umbilical

Figure 3.12 ROV and Tooling Skid

Standard ROVs require an umbilical cable to transfer the mechanical loads, power and communications to and from the vehicle. The exception to this is the autonomous underwater vehicle which can be deployed, operated and recovered without a fixed connection to the surface vessel.

A deck mounted winch is provided on the surface vessel and main umbilical utilised to deploy and recover the ROV and / or tether management system to and from the worksite. Figure 3.13 shows a workclass ROV at the commencement of deployment with the tether management system (TMS) immediately above the ROV. The TMS is lowered with the ROV by means of the main umbilical cable.

The main umbilical must be capable of lifting the weight of the ROV, any additional payload, the weight of the umbilical itself and the tether management system plus being able to withstand any dynamic loads associated with working in a subsea environment.

Steel is the most common strength-member material for main umbilicals, comprising of a steel wire galvanised externally to protect the steel from corrosion. However, synthetic fibres are also very common due to the significantly reduced weights involved and can be a suitable option for deep water operations.

The TMS is utilised to provide a protective housing for an excursion umbilical. This excursion umbilical consists of electrical power supplies and sensor cabling for both video and telemetry data and allows the ROV to be deployed to the required working depth by means of the main umbilical. The full length of the excursion umbilical can therefore be used to deploy the ROV from the TMS garage or housing to the worksite. The TMS garage remains in position with the excursion umbilical providing the ROV with the ability to be manoeuvred without any requirement to adjust the main umbilical winch. In the case of some observation class ROVs, the TMS can house the ROV as well as the excursion umbilical, which provides added protection. The ROV is housed in the TMS during deployment and recovery.

ROV Control Stations

The launch and recovery and positioning and manipulation of the ROV requires a number of controls and information displays to be provided to the ROV pilot. Dedicated ROV Support Vessels will be provided with permanent ROV Control Stations, but

Figure 3.14 Observation Class ROV with TMS Garage

for situations where temporary ROV systems are fitted onboard, temporary ROV shacks are generally deployed on the vessel deck. Whether permanent or temporary, all ROV systems require the same information to be provided and for the same systems to be available:

- Launch and Recovery Controls for the main umbilical winch, cursor wires and main lift wire (if applicable).
- Primary ROV Controls for the positioning and navigation of the ROV unit when deployed.
- ROV status display including hydraulic power unit (temperature, pressure), ROV depth, TMS and Garage information (depth of Garage, length of TMS umbilical), ROV heading and pitch and roll, thruster information and status.
- CCTV monitors and video and camera recording and visual displays, including pan and tilt information to ensure that the pilot is aware of the camera system orientation.
- Communication systems with the vessel bridge and other relevant control stations such as the launch and recovery station (if separate) and dive control if on a Dive Support Vessel.

Figure 3.15   ROV Control Station

Figure 3.16   Example ROV Pilot Console Display

Figure 3.17   Example ROV Pilot Console Display

ROV Launch and Recovery Systems

The safe deployment of the ROV from the stowed position onboard the vessel to the worksite can be performed a number of different ways, dependent on the type of system that is fitted. However, all of the systems used have the same purpose; to deploy and subsequently recover the ROV, in a safe and efficient manner, avoiding injury to the personnel involved and avoiding damage to the ROV and components and to the surface vessel. In order to provide an overview of the methods of deployment and recovery available, three types of launch and recovery systems (LARS) are examined in more detail:

- Moonpool Launch and Recovery System
- Overside Trolley Launch and Recovery System
- Overside A-Frame Launch and Recovery System

Moonpool Launch and Recovery System

The majority of purpose built ROV Support Vessels will be designed to incorporate a moonpool for ROV launch and recovery operations. The use of a centreline moonpool for these operations ensures that the ROV and ROV personnel are protected from the environment (wind, seas etc) and that the sea conditions in the splash zone will be at the minimum possible as the ROV is lowered and subsequently recovered. Operations involving the air-sea interface can potentially be the most dangerous associated with ROVs and the use of a moonpool can greatly reduce these dangers.

The reduced effects of the environmental conditions when using a moonpool LARS has the benefit of increasing the working weather envelope for the ROV systems and therefore the ROV Support Vessel. A moonpool LARS will have the following essential elements:

- Umbilical and Lift Wire Winches
- Cursor System (including Cursor Latches)

Umbilical and Lift Wire Winches

Typically, either a soft umbilical or an armoured umbilical is provided for the ROV system. In the case of a soft umbilical, the umbilical does not provide any lifting capability and therefore a separate lifting winch and lifting wire is provided for the deployment and recovery of the system. However, an armoured umbilical will provide both the main umbilical and associated services (power, camera and video feeds) and a lifting component.

Figure 3.18 Moonpool Launch and Recovery System

Cursor System (including Cursor Latches)

The use of a cursor system can significantly reduce the time period required to deploy the ROV through the air-sea interface (due to the added weight).

The cursor arrangement shown in Figure 3.18 shows the TMS arrangement enclosed by the cursor during deployment. The cursor is deployed to a position below the vessel (the cursor wires or rails are terminated at the deepest possible level on the vessel). The possibility of a pendulum effect is therefore reduced considerably, with the TMS protected within the cursor and the ROV unit itself, being held more rigidly in a vertical position.

This type of cursor is constructed of stainless steel and is hollow, in places, to allow water ingress during deployment. The added weight provided by the water

Figure 3.19 Umbilical Winch

Figure 3.20 ROV Umbilical – Armoured Exterior

ingress increases the speed of the deployment operation and therefore the time the ROV and TMS will be exposed to the splash zone. The lower parts of the cursor may be solid and filled with artificial ballast to increase the weight further.

Once the cursor has been deployed to the lowest extremity possible, it is latched in place and the deployment of the ROV and TMS continues.

Overside Trolley Launch and Recovery System

Where an additional ROV system is required to be fitted onboard a purpose built ROV Support Vessel, or where a traditional Offshore Support Vessel is modified for ROV operations, an overside trolley LARS may be used. These types of system are very similar in nature to the moonpool launched systems and will include a cursor and trolley arrangement in order to transfer the ROV from the interior of the ROV hangar, to the side of the vessel.

In the system shown in Figures 3.21 and 3.22, in the stowed position, the overhead trolley and cursor will be latched together and the trolley retracted into the ROV hangar. The ROV will be lowered to the hangar deck for maintenance and repair work. For deployment, the ROV is hoisted into the launch position, with the TMS retracted into the cursor arrangement. The trolley is then extended with the cursor, TMS and ROV. Once fully extended, the cursor is unlatched from the trolley and lowered via the vertical rails mounted on the side of the vessel. Once the cursor has been lowered to the extremity of the cursor rails, the cursor will remain in position and the deployment of the ROV and TMS continue.

This type of system provides similar protection to the moonpool launched cursor type arrangement, but does not provide the reduced movement that a centreline position or the environmental protection that an interior launch position provides.

Figure 3.21   Overside ROV Launch and Recovery System

Figure 3.22   Overside ROV Launch and Recovery System

be subject to. A latch system should be incorporated into the sheave mechanism to assist with the control of the ROV and TMS during deployment and recovery operations.

Overside A-Frame Launch and Recovery System

A-Frame launch and recovery systems provide the least protection of the three systems described here. However, on vessels where space is limited or where moonpool and overside cursor systems are already in place, an A-Frame system may be a feasible option for providing ROV support.

An A-Frame launch and recovery system is often installed on the vessel, particularly when a temporary installation, on a mounting skid arrangement. This skid arrangement can be welded directly to the vessel's deck and will include the A-Frame, hydraulic power units, umbilical winch and sheave arrangement and latch. The ROV and TMS will be integral to this mounting skid and will be stowed on the skid for maintenance and repair operations and when not deployed.

The A-Frame and sheave arrangement, together with the umbilical winch provide the means by which the ROV and TMS are swung clear of the vessels side and subsequently deployed. The winch arrangement must have suitable capacity to lift the ROV, TMS, umbilical and the appropriate dynamic forces that the system may

Figure 3.23   Overside A-Frame Launch and Recovery System

Reference

IMCA R011

The Initial and Periodic Examination, Testing and Certification of ROV Handling Systems

IMCA R011 provides guidance on the initial and periodic examination, testing and certification of ROV Handling Systems including requirements prior to delivery of the system and in service maintenance and testing regimes.

Chapter 4

OFFSHORE SUPPORT VESSEL DESIGN — CONSTRUCTION VESSELS

General Introduction

The main purpose of any construction vessel is to successfully offload and install subsea or topside hardware in an offshore environment in a controlled and safe manner. Such operations may include the use of divers or ROVs, but the major equipment utilised for construction work will obviously be the crane.

Figure 4.1 Boom Type Crane with Lattice Arrangement

In generic terms, the Construction Vessel will, in addition to the standard Offshore Support Vessel systems and equipment, have the following design features:

- The main feature on any Construction Vessel is the capability to lift and deploy subsea hardware to and from the seabed. The specification of a suitable crane is therefore the most important consideration in the design of the vessel and systems.
- Dependent on the vessel capability and capacities of the crane, the provision of a suitable anti-heeling system may be considered essential in order to maintain the positive stability and integrity of the vessel whilst heavy loads are being deployed or recovered.
- The provision of a large, open and suitably strengthened deck is essential for construction operations. In some circumstances, such as topside installations the equipment to be lifted may be mobilised to location on barges. However, for general installation operations such as subsea manifolds or spool pieces, the equipment will be loaded onto the installation vessel.
- Diver or ROV support facilities and installations.

Construction Vessels and Crane Classification

Classification Society and flag State Authority requirements and rules will be specific to each vessel and vessel type. For vessels classed by Det Norske Veritas (DNV), the crane (including pedestal and lifting gear) is not subject to class approval, unless the class notation CRANE, DSV or Crane Vessel is requested.

The certification as issued by DNV must comply with the International Labour Organization (ILO) convention No. 152. DNV has incorporated the requirements of the ILO convention into their rules for cranes.

In order to process the DNV certification, a new crane will require:

- Design Approval
- Manufacturing Survey
- Survey of the Installation Onboard
- Monitoring of Functional Tests / Load Tests

Cranes – Operational Considerations

Crane Position

The positioning of a crane will always require careful

DNV Classification Rules

Shipboard cranes are lifting appliances onboard ships and similar units intended for use within harbour areas and when at sea within the cargo deck area.

Offshore cranes are lifting appliances onboard ships and similar units intended for cargo handling outside the deck area at open sea e.g. loading and discharging of Offshore Support Vessels, barges, etc or from the seabed.

Reference

International Labour Organisation (ILO)
Convention No. 152
Occupational Safety and Health (Dock Work) Convention, 1979

The convention applies to all and any part of the work of loading or unloading any ship. The term 'lifting appliance' includes all stationary or mobile cargo handling appliances, including shore-based power-operated ramps, used on shore or on board ship for suspending, raising or lowering loads or moving them from one position to another while suspended or supported.

Reference

consideration and will depend on the nature of the operations that the vessel may be required to perform. For Offshore Support Vessels a variety of operational and safety aspects of the vessels function will have to be considered.

- In general terms, the positioning of the main deck crane has similar considerations as those involved in the positioning of a dive system or an ROV system. The positioning of the crane amidships and along the centreline of the vessel would minimise the effects on the crane and therefore the load due to the movement of the vessel. However, such positioning can considerably affect the operational capabilities of the crane. A crane situated on one side of the vessel will have a wider crane radius and therefore improved working capabilities, although only on one side of the vessel. A crane situated amidships may have limited reach over the entire after deck, whereas a crane situated midway along the afterdeck and closer to the stern will have a more general reach capability over the entire afterdeck. Additionally, a crane installed either to starboard or port, will be able to take advantage of the greater strength which generally exists in the ships hull at these points.

- For Dive Support Vessels, the crane will be expected to be utilised to assist diving operations and therefore the siting of the crane should be considered with due regard to the position of the dive system. The main crane may be required as a method for recovery of a dive bell in the event of failure of the primary or secondary handling system and this should also be considered.

- The positioning of the crane, particularly for high capacity cranes, will have considerable impact on the stability of the vessel during lifting operations. In addition, due to the heavy loads that may be deployed or recovered at the extremities of the cranes reach, the positioning of the crane will impact the level of anti-heeling system required to ensure the positive stability of the vessel.

Response Amplitude Operators

The movement of the vessel has a major impact on the operational characteristics and limitations of the vessels crane. In order to determine the theoretical behavior of the vessel when operating at sea, Response Amplitude Operator (RAO) statistics for the vessel are calculated.

The RAOs are obtained from models of proposed ship designs tested in a model basin, along with data obtained using computational flow dynamics software applied to a software model of the ship correlated with the actual physical scale model. Effectively the RAOs can indicate the effect that a specified sea state will have on the vessels motion.

Lifting Capacity and Working Radius

The purpose of the crane is to lift structures from and to the seabed from the vessel; therefore the most important aspect of the crane will be the lifting capacity and working radius, both of which are critical in determining the operational capabilities of the crane and therefore the vessel.

- The lifting capacity and working radii for internal lifts at sea, subsea lifts and ship to ship transfers should be specified at the design stage so that dynamic factors can be considered in order to achieve the required Safe Working Load for the crane.

- For offshore lifts, the vessel motions and maximum sea state for the required lifting capacity should be determined.

- For subsea lifts, the depth and the weight of wire deployed at that depth will be a factor and must be allowed for in any lifting capacity curves.

- Any capability to utilise a multiple reeve for lifts to provide a higher lifting capacity should be determined.

Figure 4.2 Lattice Boom Type Crane with Luffing Wires

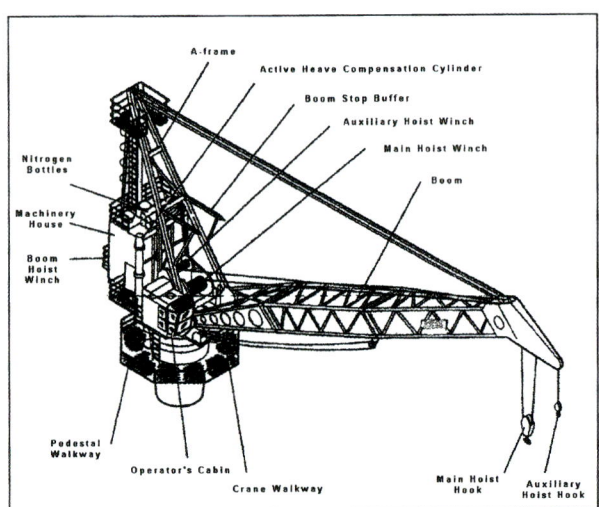

Figure 4.3 Lattice Boom Type Crane with Luffing Wires

Appendix A: Main hoist load curves

	1.76 Dynamic Factor					
	Lwire <= 740 m	Lwire <= 908 m	Lwire <= 1070 m	Lwire <= 1150 m	Lwire <= 1430 m	Lwire <= 1500 m
Radius [m]	SWL [t]	SWL [t]	SWL [t]	SWL [t]	SWL [t]	SWL [t]
4.10	100	94	90	85	81	78
4.43	100	94	90	85	81	78
4.76	100	94	90	85	81	78
5.09	100	94	90	85	81	78
5.42	100	94	90	85	81	78
5.74	100	94	90	85	81	78
6.07	100	94	90	85	81	78
6.39	100	94	90	85	81	78
6.71	100	94	90	85	81	78
7.03	100	94	90	85	81	78
7.35	100	94	90	85	81	78
7.66	100	94	90	85	81	78
7.98	100	94	90	85	81	78
8.29	100	94	90	85	81	78
8.60	100	94	90	85	81	78
8.90	100	94	90	85	81	78
9.21	100	94	90	85	81	78
9.51	100	94	90	85	81	78
9.81	100	94	90	85	81	78
10.10	99	94	90	85	81	78
10.40	96	94	90	85	81	78
10.69	94	94	90	85	81	78
10.97	91	91	90	85	81	78
11.26	89	89	89	85	81	78
11.54	87	87	87	85	81	78
11.82	85	85	85	85	81	78
12.09	83	83	83	83	81	78
12.36	81	81	81	81	81	78
12.63	79	79	79	79	79	78
12.89	78	78	78	78	78	78
13.15	76	76	76	76	76	76
13.41	75	75	75	75	75	75
13.66	73	73	73	73	73	73
13.91	72	72	72	72	72	72
14.15	71	71	71	71	71	71
14.39	69	69	69	69	69	70
14.62	68	68	68	68	68	68
14.86	67	67	67	67	67	67
15.08	66	66	66	66	66	66
15.31	65	65	65	65	65	65
15.52	64	64	64	64	64	64
15.74	64	64	64	64	64	64
15.95	63	63	63	63	63	63
16.15	62	62	62	62	62	62
16.35	61	61	61	61	61	61
16.55	60	60	60	60	60	60
16.74	60	60	60	60	60	60
16.92	59	59	59	59	59	59
17.10	58	58	58	58	58	59
17.28	58	58	58	58	58	58
17.45	57	57	57	57	57	57
17.61	57	57	57	57	57	57

Figure 4.4 Crane Capacity Curves in Tabular Format

Crane Types

There are many different crane types fitted onboard both monohull and semi-submersible crane and construction vessels. The main types associated with general offshore construction operations are the boom (or jib) and knuckle boom cranes, although telescopic cranes are also utilised in some instances.

Irrespective of the type of crane fitted, the inclusion of a pedestal is standard on Offshore Support Vessels and provides the basis on which the varieties of crane types are mounted. The pedestal type crane base consists of a steel crane housing bolted to the pedestal via a slew bearing.

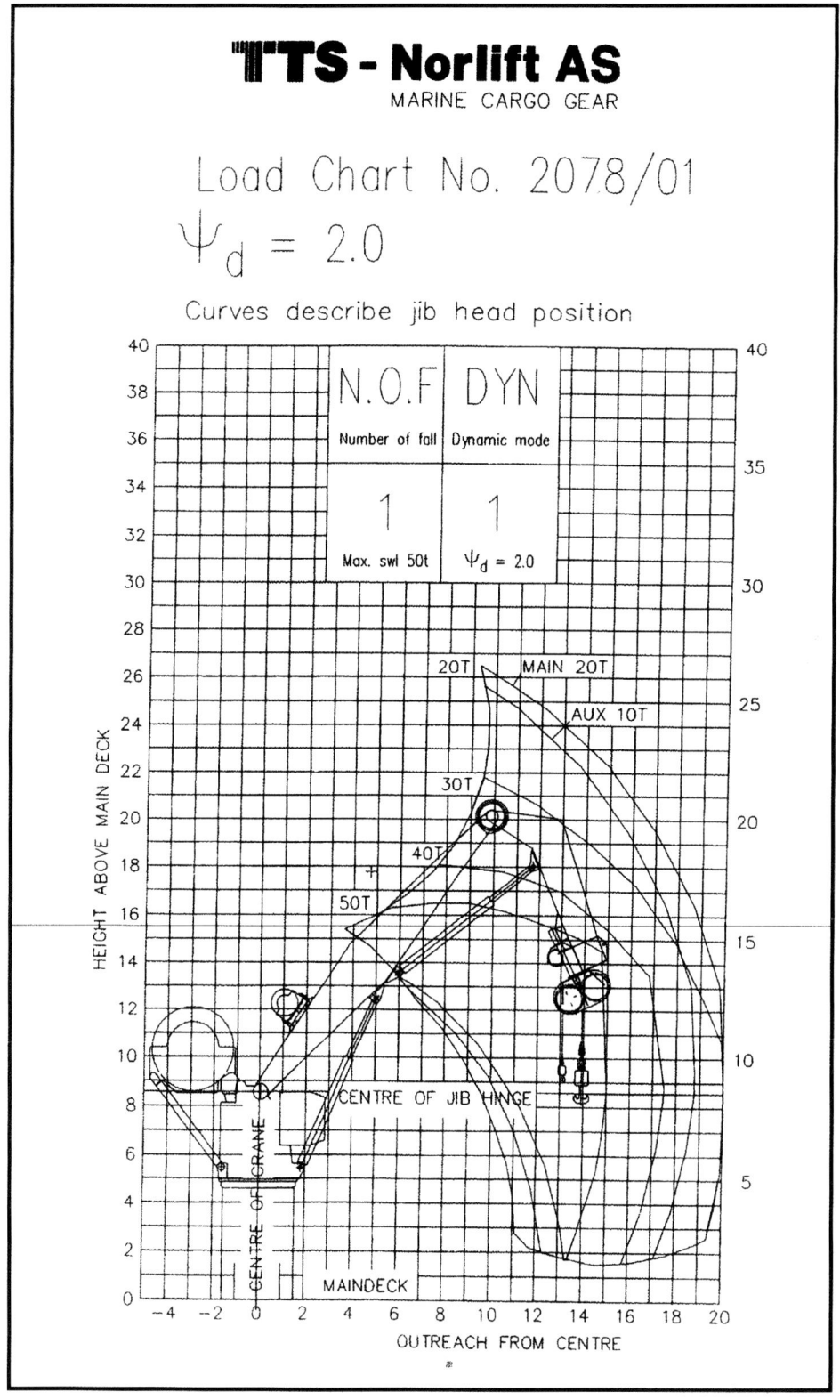

Figure 4.5 Crane Capacity Curves in Graphical Format

Boom (or Jib) Cranes

Boom type cranes can consist of a box (solid) or lattice jib arrangement with luffing cylinders or luffing wires being utilised to luff (lift and lower) the jib.

The advantages and disadvantages of the lattice boom type can include:

- Lower construction and installation costs are associated with box and lattice designs where less steel work is needed to provide the jib. Lattice booms are therefore lighter and the boom length can be modified by adding or removing modular sections.
- Less maintenance and therefore less maintenance costs are associated with box and lattice designs.

- The lattice boom design cannot withstand the same bending moments and shear forces that a box type boom can.
- The lifting point on a lattice type boom is restricted, in most cases, to the boom tip, as alternative lifting points cannot be placed along the boom length.

Figure 4.6   Lattice Boom Type Crane

Figure 4.7   Box Boom Type Crane with Luffing Cylinders

The advantages and disadvantages of the box boom type can include:

- The box type boom is more flexible than the lattice type boom and can therefore be used safely in worse environmental conditions.
- The box type boom is more likely to be fitted with a ram cylinder luffing system, rather then luffing wires. There is therefore less uncontrolled boom motion in adverse weather conditions and no requirement for luffing wires which would require regular maintenance, inspection and replacement.
- A box type boom is more resistant to damage from swinging loads due to the strength of the design.
- Due to the increased strength of the design of the box type boom, there will be the option to install additional lifting points and sheaves along the length of the boom, thereby providing the option of heavy lift rigging at reduced working radii.
- As the box type booms are generally fitted with ram cylinder luffing, a reduced working radius for the crane is expected.

Knuckle Boom Cranes

Knuckle boom (or articulated) cranes are increasingly popular onboard Offshore Support Vessels where space is limited and flexibility for a variety of operations is required.

Although the knuckle boom crane is more complex and therefore more expensive than the standard boom type crane, there are many advantages that have led to the cranes popularity.

The advantages and disadvantages of the knuckle boom type crane can include:

- The main attraction with the knuckle boom crane type is the reduction in space that is required to house the crane, and in particular the boom. When not in use the crane boom can be stowed in the folded position.
- The pendulum effects that can be induced on a suspended load due to the vessel motions can be substantially reduced as the boom tip may be positioned very close to the lifting point of the load, thereby keeping the wire length to a minimum.
- A disadvantage of the knuckle boom type crane can be that there may be some limitations when attempting to lift tall loads or loads within a short

Figure 4.8   Knuckle Boom Crane

OFFSHORE SUPPORT VESSELS   51

working radius. With such loads, the crane boom may not be the fully extended position and therefore the boom tip will be relatively close to the load.

- Maintenance of knuckle boom cranes may be more onerous due to the additional sheave arrangements and hydraulic systems required for the knuckle sections.
- The operation of knuckle boom cranes is more difficult for the crane operators due to the added complexity of the crane structure. The degree of competence, training and familiarisation for knuckle boom cranes should be specific to such cranes and their particular hazards. In addition, the crane curves for knuckle boom cranes can be more difficult to interpret.

Telescopic Cranes

Telescopic boom cranes tend to be utilised for small capacity uses such as stores and provisions cranes. This crane type has a similar advantage to that of boom type cranes, with the boom tip being close to the load, thereby reducing the pendulum effect. Space needed for telescopic cranes is obviously reduced with a large segment of the crane boom being retracted when in the stowed position. However, the weight and complexity of the crane boom are both increased due to the extending nature of the boom.

Figure 4.9 Knuckle Boom Crane in Stowed Position

Figure 4.10
Ballast transfer and stability management are essential components of crane vessel operations

Chapter 5

OFFSHORE SUPPORT VESSEL DESIGN — PIPE LAY VESSELS

General Introduction

Pipe Lay Vessels require extensive systems and equipment specific to pipe laying operations and are therefore very rarely utilised for any other purpose. In order to load out and subsequently perform pipe lay operations the following generic equipment, systems and design features will be expected on a Pipe Lay Vessel.

Figure 5.1   Skandi Navica - Pipe Lay Vessel

- Storage capacity for deck mounted or under deck carousels or reels for storage of the pipe is a major consideration. Work stations and storage facilities for pipe sections are required on traditional barge deployment systems.
- The added weight and stability considerations for deck or under deck stored pipe must be considered with the added complication of damage stability requirements for vessels with carousel holds and moonpool deployment systems.
- The deployment system for the pipe may be a stern mounted ramp system or vertical lay system via a moonpool. Both systems will require the exit route of the pipe to be clear of all possible hazards such as thrusters, propellers and hull obstructions.
- Both ramp and vertical lay systems will require equipment and systems to maintain predetermined tensions during deployment and to align and transfer the pipe from the onboard storage location to the tensioners and exit arrangement.
- Suitable crane capacity will be required for assisting with the loading and deployment of the pipe and pipeline end terminations.

Pipe Lay Vessels - Classification Society Design Principles

In order for a vessel to be given the class notation Pipe Laying Vessel under the requirements of the DNV Rules (for example), the design and construction will be reviewed in detail, with particular emphasis on the following areas of the vessel, systems and equipment:

- Design plan approval will be required for pipe support arrangements on the pipe ramp, tensioners, stingers, crane supports, stowed pipe storage areas and the reel or carousel structures. These details should also include the maximum capacities of each component and maximum weights for the stored pipe.
- Design and construction approval will be required for the hull structure relating to pipe laying operations, support structure for pipe laying equipment and for all positioning equipment relevant to pipe laying operations.
- The vessel's intact and damage stability in all operating modes must be calculated. Vessels over 80 metres in length must comply with the damage stability and subdivision requirements for cargo ships. However, the flag State Authority may also require compliance with the requirements of IMO Resolution A.534 (13) – Code of Safety for Special Purposes Ships.

Pipe Lay Vessels - Operational Design Considerations

- DP Pipe Lay Vessels which are required to operate in close proximity to Offshore Installations should have at least DP Class 2 capability.
- The thruster and propulsion units should provide good position keeping capabilities in the intact and the worst case failure conditions. Such conditions should include a subtraction of the mean tension required for working with the highest pipe tension load for which the vessel is designed.
- The DP control system software should be provided with a tension input for pipe lay operations to ensure that pipe lay initiation, lay down and recovery are

DNV Classification Rules	
Pipe Lay Vessels	
Reference	The requirements of the DNV Rules for Ships apply to vessels specially intended for laying pipelines on the seabed. Any vessel designed and constructed in accordance with the DNV requirements may be given the class notation Pipe Laying Vessel.

IMCA M103

Guidelines for the Design and Operation of Dynamically Positioned Vessels

The International Marine Contractors Association (IMCA) has issued recommendations to the industry regarding the operational design characteristics of DP Vessels, which includes Pipe Lay Vessels.

Reference

included in the DP model. It is essential that this tension input is accurate, redundant and reliable to ensure position keeping performance and stability.

- The systems and equipment onboard should be such that no single failure of the pipe lay or positioning equipment should result in the total loss of the tension in the pipe or loss of position stability. The tensioners should be designed with the same level of redundancy as the DP system and a Failure Modes and Effects Analysis should be performed on the lay system.

Pipelines and Pipe

There are a number of variations in the product that the Pipe Lay Vessel may be required to install, or in some cases, decommission:

- Rigid Steel Pipe
- Flexible Pipelines and Flowlines
- Risers
- Umbilicals

Rigid Steel Pipe

Rigid steel pipelines are the conventional method of pipeline utilised for the transportation of oil or gas products or injection of fluids along the seabed.

Flexible Pipelines and Flowlines

A flexible pipeline or flowline can be utilised for the transportation of oil or gas products or injection of fluids along the seabed. Flexible pipe can consist of a number of layers of different materials. These layers each have a specific function, such as providing insulation, strength or corrosion protection.

The use of layers within the construction of the pipes allows each pipeline or flowline to be designed for the particular purpose that it is intended. The number of layers will therefore differ dependent on the particular function of the pipeline or flowline.

In addition, flexibles can be recovered and re-used or moved to a different location.

Risers

A riser is a pipe or arrangement of flexible or rigid pipes utilised to transfer fluids from the seabed to surface installation, and for the transfer of injection or control fluids from the surface installation to the seabed.

Umbilicals

An umbilical is an arrangement of hoses and pipes and can also contain electrical or optical components for the control of subsea structures from the surface installation. The umbilical will be protected by an outer sheath or armouring.

Pipe Lay Methods

The safe and efficient deployment of a pipeline, umbilical or control line from the surface to the seabed must not compromise the structural integrity of the product. The deployment of the product in the correct location, without damage, is essentially the main purpose of the deployment vessel.

Traditional pipe lay deployment methodology consisted of separate sections or joints of pipe being welded together onboard the pipe lay vessel, prior to deployment. A longer continuous section was therefore possible to be installed on the seabed from individual pipe lengths.

Subsequent development of pipe technology has resulted in the fabrication of continuous reeled pipe which can be stored onboard the vessel and deployed in one length without the need for offshore welding. This methodology greatly reduces associated costs and increases the speed and therefore efficiency of pipe lay installation operations.

The methods available to complete pipe line installations vary from vessel to vessel, however there are four main methodologies which are described below:

- S-Lay
- J-Lay
- Reel Lay
- Vertical Lay

S-Lay

The S-lay installation methodology is the traditional method of deployment from a pipe lay barge. Pipe lengths are loaded onboard and installation operations conducted via a series a work stations on the after deck of the barge. With the pipe lengths in the horizontal, the pipe lengths are welded together, inspected and coated in each of the work stations. The welded pipe is overboarded under tension and supported by a guide arrangement (stinger). This stinger limits the overbend in the pipe by limiting the curvature of the pipe as it exits the vessel and the tensioners prevent pipe damage due to buckling. The size of the stinger arrangement is directly related to the pipe diameter, coating thickness and water depth.

As the pipe is overboarded, the pipe lay barge is moved ahead and the sequence is repeated.

The S-lay methodology is characterised by the double bend in the pipe as it is released to the sea, forming an 'S' track.

J-Lay

The J-lay installation methodology utilises a vertical or near vertical tower structure. The pipe lengths are lifted from the storage location on deck, pre-jointed into triple or quadruple joints and placed into the tower structure where welding, inspection and coating is completed.

The use of this methodology eliminates the potential for overbending of the pipe which can occur in S-lay operations, as the pipe exits the vessel in a near vertical aspect. Laying the pipe at near-vertical angles also reduces the distance to the touchdown point.

The main advantage of the J-lay method in comparison with S-lay is for deepwater installations where the near vertical deployment minimises tensions and allows the installation to be under more control. The tensions associated with the S-lay method become unfeasible for deeper water installations.

The J-lay methodology is associated with vertical deployment and characterised by a single bend in the pipe just above the seabed, forming a 'J' track.

Reel Lay

The reel lay installation methodology can be utilised to deploy continuous lengths of pipe. The pipe is loaded onboard the vessel on deck or under deck mounted reels or carousels and minimises the need for welding offshore, allowing increased deployment speeds.

Offshore, the pipe is spooled off from its storage location, via straighteners and tensioners to the seabed, as the vessel moves ahead. The reel lay method can include both J-lay and S-lay operations dependent on the arrangement of the straighteners, tensioners and associated equipment.

Figure 5.2   Reel Lay and Ramp Arrangement

Vertical Lay

The vertical lay installation methodology is a variation on reel lay and J-lay, but with a fixed vertical departure angle of the product. A vertical lay system is ideal for the deployment of flexibles, umbilicals, control lines and any products with a relatively small bending radius.

Offshore, the product is spooled from its storage location into the vertical lay tower and is deployed vertically via a moonpool. The vertical lay method, when utilised with the vessel moving ahead during deployment, will result in a similar deployment track to J-lay.

Example Pipe Lay Systems

The systems and equipment and their layout onboard Pipe Lay Vessels varies widely from vessel to vessel and can result in very little compatibility or similarity between systems. The variation in the storage methods utilised for the product (deck reels, under deck or deck mounted carousels) and the deployment methods (stern ramps, side mounted chutes and vertical towers) can, with certain exceptions, be combined to form unique lay systems.

For this reason it is very difficult to describe a generic pipe lay system, as each and every system will be unique in some way. However, in order to provide an introduction and brief overview of the type of systems currently in operation, two lay systems will be described.

- Skandi Navica – Deck Mounted Reel with Stern Ramp Arrangement
- Toisa Perseus – Deck Mounted Reel or Under Deck Carousels with Vertical Lay Tower.

Skandi Navica – Deck Mounted Reel with Stern Ramp Arrangement

The pipe lay systems and equipment onboard the *Skandi Navica* can be divided into the following

OFFSHORE SUPPORT VESSELS   55

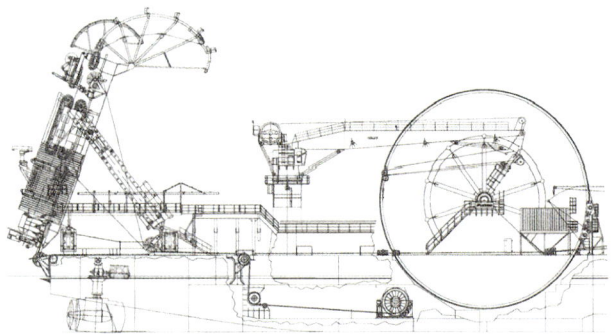

*Figure 5.3   Skandi Navica
Deck Reel and Stern Ramp Deployment Arrangement*

component parts:
- Deck Mounted Reel
- Ramp Fleeting and Elevation System
- Aligners and Straighteners
- Tensioners
- Pipe Clamp Arrangement
- Exit Monitoring Frame
- Exit Roller Box
- Abandonment and Recovery Arrangements
- PLET Handling Frame

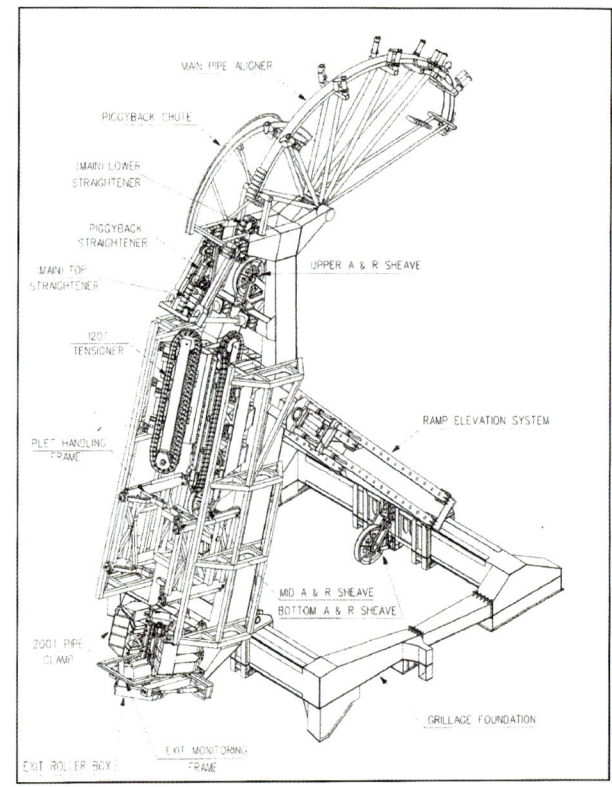

Figure 5.4   Ramp Arrangement

Deck Mounted Reel

The function of the reel is to store the pipe for transit and subsequent deployment at the offshore location. The reel is driven by hydraulic motors and therefore loading of the pipe is controlled under power of the reel and hold back tension (by the tensioners) is provided in order to control the curvature of the pipe during pipe lay operations, between the reel and the ramp arrangement.

The main deck reel has a drum diameter of 15 metres, a flange diameter of 25 metres and a drum width of 6.7 metres which equates to a potential pipe loading capacity of 2500 tonnes. The reel is located on the centreline of the vessel and midships fore and aft for stability reasons. Each pipe loading operation will be calculated on a project by project basis; however the table in figure 5.8 will give an indication of the scale of the main deck reel and the range of capacities that the reel can hold.

A tie-in arrangement is provided for the storage of the pipe end connection in a protected drum compartment. The pipe end connection is attached to the drum via a strop and shackle arrangement.

Ramp Fleeting and Elevation System

The function of the pipe lay ramp arrangement is to house all the components required to deploy the

Figure 5.5   Ramp Arrangement

Figure 5.6 Deck Mounted Reel - External

Figure 5.7 Deck Mounted Reel - Internal

Figure 5.9 Deck Mounted Reel – Tie-in Arrangement

Figure 5.10 Ramp Arrangement

pipe from the storage reel to the seabed. This type of arrangement allows the entire ramp to be elevated, changing the departure angle of the pipe from the vessel and also allows the ramp to be fleeted (moved to port or starboard). The fleeting of the ramp ensures that the pipe can be transferred directly from any section of the reel drum wrap to the pipe tensioners without any lateral bending of the pipe. The ramp is therefore fleeted as the pipe is unspooled from the reel and the departure point from the reel changes.

As the stern ramp angle is adjustable between approximately 20 and 90 degrees, the product can be loaded directly from the quayside onto the storage drum.

The ramp assembly supports the aligners, straighteners, tensioners, PLET handling frame, pipe clamp, exit roller box and exit monitoring frame.

Aligners / Straighteners

The pipe aligners are attached to the top of the ramp arrangement and are utilised to guide the pipe from the deck mounted storage reel to the straighteners. The pipe exits the straighteners to the tensioners aligned and centred.

Pipe capacities						
Pipe diameter (inch)	6	8	10	12	14	16
Minimum wall thickness (mm)	5.7	8.5	11.6	15.0	17.4	22.0
Storage capacity (km)	69	40	27	19	16	12
Weight of pipe (tonnes)	1690	1840	2100	2200	2353	2500

Figure 5.8 Deck Mounted Reel Capacities

The aligner consists of a fabricated radius arm which provides support of the pipe as it is spooled on or off the main reel. A hydraulically powered aligner chain track assembly is fitted to the radius arm and is made up of a series of rollers and chain pads.

Figure 5.11   Main and Piggy Back Aligners

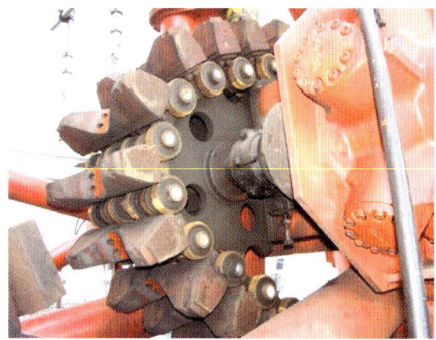

Figure 5.12   Main Aligner Track

A piggy back chute is provided alongside the main pipe aligner for the alignment and support of small diameter flowlines or umbilicals from the piggy back reel. This piggy back chute is not powered, but is coated with nylon to protect the product during loading and deployment operations.

The main straightener is positioned between the aligner and the tensioners. The straightener track is utilised to apply a reverse curvature to the pipe and therefore deliver the pipe to the tensioner in a fixed aspect. The straightener track can be adjusted to pre-defined settings, which are calculated prior to loading and proven during subsequent straightening trials. The piggy back straightener consists of adjustable rollers which can be similarly set to specific requirements.

Figure 5.13   Piggy Back Straightener

Tensioners

The principle function of the tensioner arrangement on any pipe lay spread is to provide tension assist and control pipe movement during spooling and deployment operations.

The configuration of the tensioner arrangement is formed on two basic concepts:

- Linear tensioners grip the pipe collars with a clamp and then lower the pipe a set distance. A second clamp is moved to a position above the first clamp and the process repeated.
- Tracked tensioners consist of a number of continuously revolving tensioner tracks which hold the pipe in position. This is the system utilised onboard *Skandi Navica*.

The tracked tensioner method is the more common system in use and can consist of between two and four tracks. For rigid pipe lay operations from a storage reel or carousel, this type of tensioner has significant time saving advantages and is more suited to the deployment of sensitive products such as control lines and umbilicals where the clamp method may pose a potential for damage due to crushing.

The tracked type tensioner can be hydraulically adjusted dependent on the size of product with the squeeze pressure that is applied to maintain the product tension, based on calculated pipeline analysis. Each track is mounted on moveable guide rails and can be clamped and retracted as required. The tracks are fitted with tensioner pads, which can be changed out dependent on the pipe diameter.

To illustrate the capacity of these tensioners, the *Skandi Navica* arrangement can, dependent on the particular product, maintain a line tension of up to 158 tonnes for 6 inch to 16 inch diameter pipe.

The main factors determining the dimensions of a tensioner arrangement will include the total tension to be exerted on the product, the maximum clamping force, the angle of contact between the product and the tensioner pads and the coefficient of friction between the product and the pads.

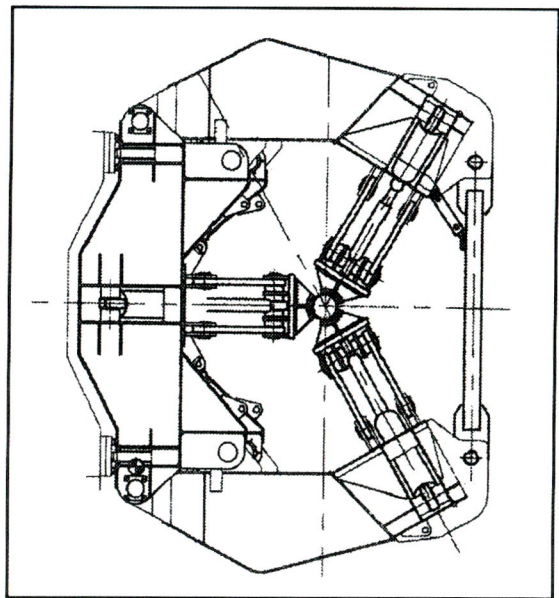

Figure 5.15   Pipe Clamp Arrangement

Exit Monitoring Frame

The pipe exit monitoring frame is positioned aft of the pipe clamp. The system consists of two arms fitted with sensors that measure the pipeline lift and are used as a guide to maintain the predetermined pipe laying parameters. The system is hinged so that the arms can be opened to allow the passage of pipeline end terminations and fittings.

Figure 5.14   Tracked Tensioner Arrangement

The total force required to hold a pipe during lay operations is governed by a number of aspects, including:

- The submerged weight of the pipe.
- The pipe departure angle and the catenary of the suspended pipe.
- The dynamics of the pipe and the vessel.
- Contingency arrangements, including the provision for the recovery of a flooded pipe from the seabed.
- Coefficient of friction between the tensioner pad and product coating.

Pipe Clamp Arrangement

The pipe clamp arrangement consists of three pad inserts (neoprene or steel lined), which are hydraulically operated and utilised to hold the pipe during welding, deployment, recovery and abandonment.

The clamp can be used for the variation of pipe diameters that the lay system is designed to deploy and can be modified to accommodate larger flexible sections and fittings.

Figure 5.16   Exit Monitoring Frame

Exit Roller Box

The exit roller box is the departure point for the pipe from the Pipe Lay Vessel and consists of a series of roller fairleads which provide protection to the product horizontally and vertically. The use of this exit roller box prevents the product being subject to damage from contact with any other part of the vessel at the departure point.

Figure 5.17   Exit Roller Box

Figure 5.19   PLET Handling Frame for Deployment of an End Termination

Abandonment and Recovery Arrangements

Winch arrangements are provided for the deployment of the product and also for situations where there may be a need to abandon a pipe lay operation and subsequently recover the pipe for recommencement of deployment operations. A series of sheaves within the pipe lay ramp assembly allow the abandonment and recovery winch wire to be connected to the pipe end to allow the pipeline to be deployed or abandoned.

Toisa Perseus – Deck Mounted Reel or Under Deck Carousels with Vertical Lay Tower

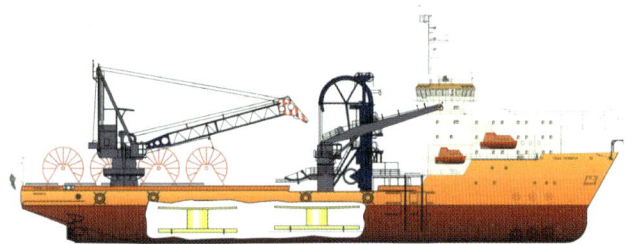

Figure 5.20   Toisa Perseus – Vertical Lay Tower and Under Deck Carousels

The pipe lay systems and equipment onboard the *Toisa Perseus* can be divided into the following component parts:

- Under Deck Carousels
- Deck Mounted Reels
- Loading Tower
- Deck Radius Controller
- Vertical Lay System Tower

These systems can be used in a variety of configurations, allowing both under deck and deck stored products to be deployed using the vertical tower and to be loaded onboard via the loading tower.

Under Deck Carousels

The *Toisa Perseus* is fitted with two under deck carousels, both of which can be loaded from deck or shore mounted reels, via the loading tower and spooling arm, and which can be used to deploy product, via the deck radius controller and vertical lay system tower. To increase the range of the products that such under deck carousels can be utilised to store, it is often possible to fit radius inserts or false cores, which increase the inner

Figure 5.18   Abandonment and Recovery Sheave at the Base of the Tensioners

PLET Handling Frame

In addition to the deployment and installation of the pipe, the Pipe Lay Vessel will also be required to deploy the Pipe Line End Terminations (PLETs). In this ramp arrangement, a PLET handling frame is an optional system, fitted below the ramp assembly for the deployment of the PLETs.

carousel drum diameter, thereby allowing products with a larger minimum bending radius to be stored.

The table below provides an approximate estimate of the product lengths that can be stored onboard each under deck carousel onboard the vessel, which will provide an indication of the size of these carousels.

Product diameter	Storage length (m)
100	80,842
150	35,930
200	21,210
300	8,982
350	6,599
400	5,053
450	3,992

Figure 5.21 Under Deck Reel Capacities

Figure 5.22 Under Deck Carousel Spooler and Carousel

Figure 5.23 Loading Product onto Under Deck Carousel

Deck Mounted Reels

Deck mounted vertical reels or horizontal carousels are a further option onboard Pipe Lay Vessels with vertical lay system towers, such as the *Toisa Perseus*. Product loaded on such deck reels or carousels can be deployed, via the deck radius controller and vertical lay tower. Overboarding chutes will be provided.

Figure 5.24 Deck Mounted Carousel

Loading Tower

The primary function of the loading tower is to guide and provide a back-tension when loading flexible product from the quayside onto the vessel. However, the loading tower can also be used to deploy product from the under deck storage carousels and for recovery of flexible from the sea bed. The loading tower consists of a single track tensioner and lower chain drive. The back-tension or deployment tension is provided by the squeeze pressure between the track tensioner and chain drive. Adjustable guide pads, attached to the loading tower structure, guide the flexible onto the lower chain track.

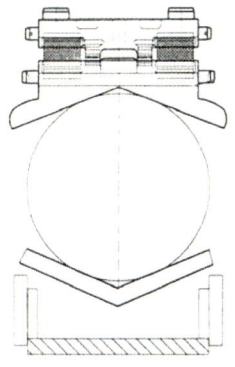

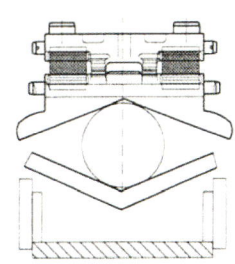

MAXIMUM DIAMETER   **MINIMUM DIAMETER**

Figure 5.25 Minimum and Maximum Squeeze Pressure Arrangement on Loading Tower

Figure 5.26 Loading Tower and Deck Radius Controller

Figure 5.27 Loading Tower Tensioner Track

Figure 5.28 Deck Radius Controller

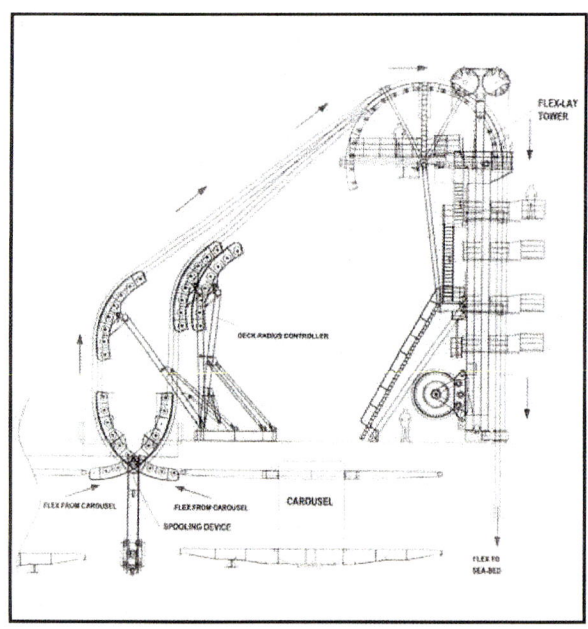

Figure 5.29 Deck Radius Controller

Deck Radius Controller

The deck radius controller is utilised to guide the product from the under deck carousels and carousel spooling device to the vertical lay system tower and the positioning of the unit can be adjusted hydraulically dependent on which under deck carousel is being used for the product deployment

The main arm consists of a series of rollers and guides to lead the product to the upper chute arrangement on the vertical lay system tower.

Vertical Lay System Tower

The vertical lay system tower consists of a lower and upper chute arrangement, tensioners, hang off clamp, moonpool and abandonment and recovery arrangement.

The lower chute is utilised to guide the product from deck reels onto the upper chute arrangement and the vertical lay system tower. The upper chute can be used to guide the product from deck mounted reel drives and also from the under deck carousels via the deck radius controller.

Figure 5.30   Vertical Lay System Tower – Upper Chute

Tensioners are provided in two vertical banks of four tracks, which can also be operated in a two track mode. As with a reel lay ramp system, the tensioners provide the tension for deployment operations and are fitted with pads appropriate to the product being deployed.

Figure 5.31   Track Tensioners

A hang off clamp is provided at the base of the vertical lay system tower, directly above the moonpool. This hang off clamp is used to hold the product during the attachment of buoyancy, during end termination operations, jointing and for abandonment and recovery.

The moonpool through which the product is deployed is at the base of the vertical lay system tower and can be sealed off when deck operations are required within the tower (as shown in figure 5.33).

Figure 5.32   Vertical Lay System Tower with Track Tensioners

Abandonment and Recovery Arrangements

Winch arrangements are provided for the deployment of the product and also for situations where there may be a need to abandon a pipe lay operation and subsequently recover the pipe for re-commencement of deployment operations. A series of sheaves within the pipe lay system allow the abandonment and recovery winch wire to be connected to the pipe end to allow the pipeline to be deployed or abandoned.

Figure 5.33   Hang Off Clamp

Chapter 6

DIVING OPERATIONS

General Introduction

Diving operations can be considered the most critical operation that any Offshore Support Vessel will be involved in. The introduction of human life into the subsea working environment poses many additional risks and as such, diving operations must be carefully planned, implemented, managed and controlled.

To conduct a diving operation safely, many factors must be considered which are specific to the diving operation, diving vessel and the specific location of the operation. For the purposes of this section, the specific regulations and guidance applicable in United Kingdom waters will be considered and specific safety and operational factors identified.

- Diving at Work Regulations 1997
- Diving Operations
- Dive Planning
- Diving Operations from Vessels Operating in Dynamically Positioned Mode
- Umbilical Management
- Diving Operations within Anchor Patterns
- Diving in the Vicinity of Pipelines and Wellheads
- Diving Operations associated with Lifting Operations
- Hyperbaric Evacuation

Diving at Work Regulations 1997

The Diving at Work Regulations 1997, issued by the Health and Safety Executive (HSE), impose strict requirements and duties on the diving contractor for all diving related activities. The methods that the diving contractor utilises to fulfil these requirements and duties are not prescribed. However, in all cases the basic need to maintain the health and safety of the divers and all associated personnel is the primary purpose. The Merchant Shipping (Diving Safety) Regulations 2002 impose similar requirements.

To comply with the Diving at Work Regulations, the diving contractor must:

- Conduct an assessment of the risks to the heath and safety of all personnel involved in the diving operation.
- Prepare a diving project plan taking into consideration any of the risks highlighted in the risk assessment. This plan must be continuously updated and amended during the diving operation ensuring that any changes to the proposed plan are suitably assessed.
- Appoint a person to supervise the diving operation. It is the responsibility of the supervisor to ensure that all personnel involved in the diving operation are fully aware of the diving project plan. The appointment of the diving supervisor must be confirmed in writing by the diving contractor.
- Provide suitable and sufficient resources in the form of divers and associated support personnel to conduct the diving operation in a safe manner, including emergency situations and situations where first aid assistance to the divers may be required.
- The diving contractor must ensure that all personnel are suitably certified and competent to conduct the tasks for which they are employed. These personnel must comply with the requirements and prohibitions imposed under the Diving at Work Regulations and provisions of the diving project plan.
- Provide suitable and sufficient resources in the form of equipment and systems required to conduct the diving operations in a safe manner, including emergency situations and situations where first aid assistance to the divers may be required.
- Maintains the diving systems and equipment and all associated equipment in a safe working condition.
- Maintains records containing the particulars of each diving operation. Such records must be maintained for at least two years.

In addition to the requirements imposed on the diving contractor, there are considerable duties that are placed on the divers themselves. These duties require that the divers must be in possession of an approved qualification, valid certificate of medical fitness to dive

Reference

Diving at Work Regulations 1997

Divers are legally required to have in their possession offshore a valid diving medical certificate. This certificate must be signed by a doctor approved by the Health and Safety Executive according to the criteria as specified in Regulation 15 of the Diving at Work Regulations 1997 and part 2 of the Health and Safety at Work Act 1974.

and that they must retain diving records for a period of at least two years.

Diving Operations

Diving operations can include many diverse tasks including construction support, inspection and repair and maintenance tasks.

- Construction Support – The installation and eventual de-commissioning of offshore structures including platforms, subsea manifolds, pipelines and floating production, storage and offloading (FPSO) installations may require diver intervention. This diver intervention may include assistance in positioning of the installation components, manipulation of valves and connection and disconnection of lifts subsea.
- Inspection – The inspection of offshore structures can include subsea installations such as pipelines and manifolds, but also the underwater components of surface structures such as platforms and drilling rigs. Inspection tasks can include integrity checks (MPI, ultrasonic, NDT), marine growth surveys and cleaning operations, damage and condition surveys, debris inspection and removal and scour surveys).
- Repair and Maintenance – Repair and maintenance operations can include hyperbaric welding, equipment (such as valve) replacement, anode replacement, sandbagging of pipelines, installation of mats, dredging operations and integrity checks on equipment (bolt tightness checking).

Dive Planning

The detailed planning and preparation of a diving operation should consider the divers environment, the diving vessel, the diving equipment and systems, the hazards and potential risks to the divers and also any potential emergency situations and contingencies. The planning should not, however, be left to the diving contractor alone. The design and construction of offshore surface and sub-surface installations and structures should follow some basic design philosophies. The need for any underwater human or mechanical intervention within the periphery or any need for deep penetration of the structure should be avoided and any attachments used only for the fabrication and construction phases should be removed prior to installation. These simple design features will significantly reduce the hazards to divers from potential traps and snags that could be life threatening to the diver.

The evaluation and assessment of the diving operations and the dive project plan should therefore consider, at a minimum, the following:

- The location of the diving operation with due regard to any identified pipelines, cables, debris or any other obstructions that may be a hazard to the divers.
- The predicted and immediate environmental conditions at the location including the surface sea state, surface and sub-surface visibility and tidal and current conditions.
- The water depth at the location and the decompression requirements for the divers.
- The capabilities and limitations of the Dive Support Vessel being utilised. The position keeping and manoeuvring characteristics and redundancy of the vessels thrusters and main propulsion will be of particular importance. The Dynamic Positioning System Classification and availability of reference systems must be considered.

Figure 6.1 Preparation for Diving Operations

- The proposed task including the planned duration of the operation and any restrictions this may impose on the diving operations.
- The vicinity of other vessels and other operations in the immediate vicinity.
- Any simultaneous operations being conducted, such as crane or ROV operations where there may be a potential safety hazard from crane wires, lift (air) bags, ROVs or taut wires.
- Umbilical Management in relation to the proposed

IMCA D014

International Code of Practice for Offshore Diving

The IMCA International Code of Practice for Offshore Diving provides guidance for the completion of diving operations in a safe and efficient manner. These guidelines do not take precedence over National Regulations, but may be considered as good practice in areas where no such statutory requirements are in place.

Reference

operation, location of the work site in relation to the support vessel, diving bell and umbilical tether may provide particular hazards and restrictions on the diving operation. The length of the diver's umbilical must be considered in respect of the water depth and distances to any potential hazards such as the support vessel's thrusters and main propulsion systems.

- For diving operations in the vicinity of surface structures such as platforms, the position and use of discharges and intakes, umbilicals, drill strings and other potential hazards.

- Any operations being conducted on the surface structure such as pressure testing, crane operations or over side work which could result in the potential for dropped objects or excursion into the diver's environment.

- Diving operations may include the need for specialised equipment such as jetting, grit cleaning or burning apparatus. Specific hazards associated with the specialised equipment should be considered and evaluated.

- Underwater electrical equipment such as anodes may require isolation.

- Communication between dive control and the installation is essential. Permits must be in place when working in close proximity of installations to ensure that work is co-ordinated in a safe manner.

- In addition to the operational hazards that the divers may be exposed to during diving operations, there are physical and physiological considerations which pose serious hazards to their health and safety. Decompression sickness poses a significant hazard.

- There must be a direct communication link between the diving supervisor and the ROV supervisor (or pilot). In addition, the diving supervisor must be supplied with a repeater monitor showing the same picture as seen by the ROV pilot.

- The ROV deployment site must be a reasonable distance away from the diving bell, diving basket or taut wire launch positions in order to minimise chances of umbilical entanglement.

Diving Operations from Vessels Operating in Dynamically Positioned Mode

The stability of the Dive Support Vessel is essential in order to provide the divers with a safe working environment when deployed sub-surface. Dynamic positioning of the surface vessel is an integral part of the diving operation and therefore close co-operation between the dive supervisors and the bridge team is essential. The positioning and movement of the vessel whilst divers are being deployed, recovered or during diving operations must be strictly controlled and all dynamic positioning operations should be conducted with due regard to the position and safety of the divers and diving bells. (See also The DP Operator's Handbook, also published by The Nautical Institute).

The risk assessment and dive project plan must therefore consider the dynamic positioning operations and any planned or unplanned movement of the vessel. The following factors should, as a minimum, be considered in the planning and risk assessment for diving operations and for vessel movements when divers are deployed:

- Communication between the dynamic positioning operators, dive supervisors and the divers is essential. Any changes to the vessel's status and any movements must be discussed and agreed before execution. Any major changes in the vessel status (such as start up of another consumer) will require the divers to return to the diving bell.

- Safe working limits should consider the time required for divers to return to the bell in the case of a yellow or red alert. Such limits will be dependent on the location (open water or internal structure), water depth and any considerations specific to the operation.

- Diving bell and diver's umbilicals pose particular fouling hazards with the vessel's propellers, thrusters and other structures and obstructions. Such hazards must be considered during diving operations and vessel movements.

- Vessel movements should only consist of a heading change or a position change. A combination of both movements should not be carried out. All such movements should be conducted at slow speed, in incremental steps and should include settling periods between each step. Where heading changes are to be made, due consideration should be taken of the centre of rotation of the vessel.

- Vessel movements should not be made in joystick or manual control, except for emergency situations.

Information

Decompression Sickness

Decompression tables are used to calculate the duration of decompression that is required for particular dives. Such tables cannot completely negate the potential for decompression sickness as many risk factors such as the age and general health of the diver and dehydration levels may all have an effect on the diver's wellbeing.

>
> **Reference**
>
> **IMCA M103**
>
> **Guidelines for the Design and Operation of Dynamically Positioned Vessels**
>
> The IMCA guidelines advises that a system of visual and audible alarms be provided in Dive Control, Saturation Control, Air Diving Control, Working Area, Engine Control Room and where applicable the ROV Control Position. These alerts should be manually activated from the DP Control Station and consist of green, yellow and red signal alarms.
>
> Green Light — A steady green light indicates the vessel is under automatic DP Control. Full diving operations can be undertaken.
>
> Yellow Light — A flashing yellow light indicates a degraded DP situation. Divers should suspend operations and move to a safe location. The divers returning to the bell and subsequently the surface may be an option.
>
> Red Light — A flashing red light indicates an emergency DP situation. The divers should immediately return to the diving bell and be recovered to the surface as soon as possible. The red alert should, additionally, sound in the cabins of critical personnel such as the Master, Offshore Manager and Diving Supervisor.

- Vessel movements should not take place unless a minimum of three position reference systems are available and the movement should not exceed the scope of any of these reference systems.
- The dive supervisor and divers must be fully briefed on the movement of the vessel before commencement and may return to the diving bell. Similarly, prior to the deployment of any subsea equipment (ROVs, cranes, taut wires), permission must be obtained from the diving supervisor and the divers may be returned to the diving bell.
- If there are any doubts concerning diving operations or vessel movements, the move should be stopped and risks and control measures re-assessed.

Umbilical Management

The main umbilical from the support vessel to the diving bell and the diver's excursion umbilical from the diving bell to the diver are the diver's lifelines. Loss of either umbilical may have catastrophic results and therefore umbilical management must be strictly controlled.

A diagram specific to each vessel should be provided at both the DP control and dive control locations. Ideally, the diagram should include:

- Positions of the diving bell launch and recovery stations and the diving bell in 10 metre incremental positions with distances to the nearest thruster or propulsion system.
- Positions of the thrusters, propulsion systems and other potential hazards to the main and excursion umbilicals, such as hull obstructions or sea water intakes.
- Positions of any onboard equipment and systems that may impact diver operations, such as cranes, taut wires and ROV launch and recovery stations.

In addition to static diagrammatic information, 'live' displays of the vessel's position, subsea and surface structures, hazards and obstructions, mooring lines, diving bell location and the position of ground based reference systems (beacons) should be provided.

The information provided in diagrammatic form and in 'live' displays is utilised to ensure that the diving bell, divers and umbilicals are maintained in a safe controlled environment. Physical restrictions are also placed on the umbilicals in order to ensure that the diving bell and divers cannot be inadvertently deployed to a position of danger.

- The diver's excursion umbilical should be kept at a minimum to prevent the potential for snagging and must be physically restrained to prevent it from coming within five metres of hazards such as thrusters and propulsion systems.
- The length of the standby diver's umbilical should be two metres greater than that of the working diver's umbilical. This will ensure that the standby diver has added manoeuvrability in emergency situations, but should be similarly restrained to prevent it from coming within three metres of hazards such as thrusters and propulsion systems.
- The diver's excursion umbilical should be clearly marked along the entire length to allow monitoring of the length deployed.
- The full excursion umbilical should not be deployed, allowing a reserve for emergency situations.

Diving Operations within Anchor Patterns

Diving operations within an anchor pattern of another vessel or semi-submersible introduce particular dangers associated with the movement of the third party vessel, the movement of the vessels moorings, the interaction between the two surface vessels and the potential for contact or impedance of the diving bell, umbilicals or divers with the moorings.

Figure 6.2 Diver's Excursion Umbilical

Figure 6.3 Diver at Work Subsea

In most cases, such as a semi-submersible, the surface structure will be moored in a static location and movement of the unit will be limited. However, surface structures such as floating production storage and offloading (FPSO) vessels may be moored around a rotating turret and the vessel itself will be prone to weather vaning, where the vessel will swing with tide, current and wind and sea conditions.

- The position of the third party vessel's anchors, mooring line lengths deployed and mooring line tensions must be known in order to determine the catenary of the mooring line. For diving operations the position of the mooring lines in relation to the Dive Support Vessel, diving bell and divers should be confirmed by at least two means, one of which can be calculation.

A horizontal clearance of at least fifty metres should be maintained between the suspended mooring line and the diving bell. If the clearance is less than fifty metres, then the position of the mooring line must be monitored throughout the operation by means of, for example, an ROV mounted transponder and the operations restricted to the minimum period possible.

- For many operations, it may be decided during the risk assessment and planning stages, that the mooring line can be safely lowered to the seabed in order to reduce the hazards to the divers. However, in such circumstances the successful completion of the mooring line tension reduction should be confirmed by either visual inspection (ROV) or by confirmation of the mooring line tension by the third party vessel.

- During diving operations, the tensions in the mooring lines should be maintained in order to maintain the catenary and position of the moorings. Any movement of the mooring lines will pose a potential hazard for the divers who are deployed to the worksite.

- Communications between the moored vessel and the Dive Support Vessel must be maintained at all times. Any changes to the position or aspect of the moored vessel or the mooring line tensions must be communicated to the Dive Support Vessel.

- The interface between the moored vessel and the Dive Support Vessel must include cross referencing of the permit to work for diving operations. Onboard the moored vessel, all parties must be aware of the diving operations and controls put in place to ensure that the divers are not subject to potential hazards from overside working, adjustments to mooring line tensions or any other activity that could impinge on the diving operations.

- Prior to any extended operations, particularly involving diver intervention, the environmental limits should be discussed and agreed between the moored vessel and the Dive Support Vessel. These limits will generally be determined by the limitations imposed on the divers, but must be agreed prior to commencement of operations.

- Position and navigation data should be provided to the Dive Support Vessel bridge and dive control teams, detailing the positions of both vessels and if possible, the positions of each mooring line. Catenary curves for the mooring lines should be available to determine the calculated distances between the diving bell, divers and the mooring lines.

- The manoeuvrability of the Dive Support Vessel may be restricted when working in close proximity of anchor mooring lines. Due consideration at the planning and risk assessment stages should therefore be given to emergency and contingency situations and the escape routes that may be available for the Dive Support Vessel. Such emergency and contingency plans should take into consideration the movement of the moored vessel (FPSO rotation) and the prevailing and forecast environmental conditions.

Diving in the vicinity of Pipelines and Wellheads

Subsea operations in the vicinity of pipelines under pressure and wellheads should be conducted, if possible, by the use of ROVs. However, there may be instances where access by ROV is not possible and in such circumstances, diver intervention may be necessary. In the event of diver intervention, the main concern will be the reduction of the pipeline pressure to a safe level prior to deployment of the divers.

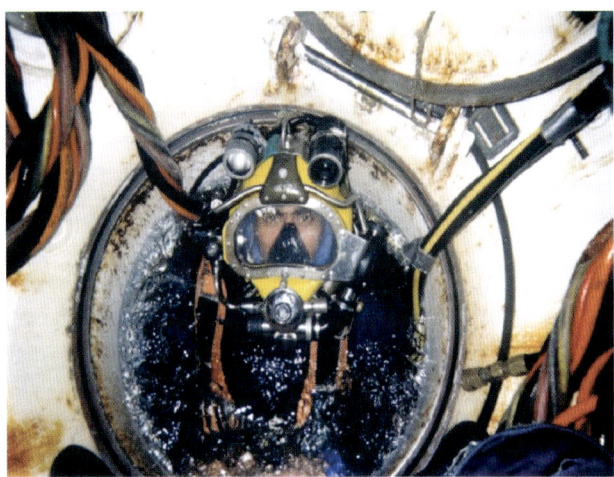

Figure 6.4   Diver exiting the Dive Bell

For diver intervention in the vicinity of pipelines or wellheads, including pipeline testing, well intervention work, flow base installation, umbilical tie-ins and control module installations, the following should be considered:

- Communications between the Dive Support Vessel and the platform control room must be maintained at all times and the interface between both sites must ensure that the platform permit to work takes into consideration the diving operations. For example, pumping equipment on the platform may need to be isolated or pressure settings limited and warning notices should be posted to this effect, access to associated pipeline equipment restricted and all relevant personnel advised of the ongoing operations.
- The risk assessment and planning stage should consider any potential hazards that may be associated with diving operations in the vicinity of a wellhead. Such hazards may include entrapment or entanglement of the divers or their umbilicals, dropped objects from the platform, exhausts and discharges, overpressure of pipelines and valves, hydrocarbon releases, use of chemicals and any galvanic protection systems. Diving in areas of restricted movement or semi enclosed or confined spaces should be considered, in particular when associated with the potential release of gases such as hydrocarbons.

- The positioning of the Dive Support Vessel and hence the diving bell, should be discussed at the planning and risk assessment stage and should take into consideration the sudden release of gas from the pipeline. The position of the Dive Support Vessel and the diving bell, in such circumstances, should not result in bell contamination or instability. Consideration of the environmental forces in the vicinity should be included in any such assessment.
- If a leak test is being conducted, the chemical utilised for such a test should be considered with due regard to the potential harm such a chemical could pose to the divers or the diving bell atmosphere.
- If the operations involve the installation of any subsea equipment, such as mattresses, lifting operations should be conducted with due regard to the position of the pipelines and the potential for dropped objects and dropped object excursions.

Diving Operations associated with Lifting Operations

Lifting operations whilst divers are in the water can pose a considerable hazard. The potential for dropped objects is the main threat; however the movement of loads in a subsea environment can also be a major consideration.

- Communications between the crane operator and dive control must be maintained in addition to the standard communication lines for normal operations to ensure that the divers are fully aware of all planned and in progress crane operations.
- The intended load path from the storage position on the deck of the vessel to the intended landing location on the seabed should be carefully planned to ensure that the load does not pose a potential hazard to the diving bell and divers when deployed. The landing position and immediate seabed vicinity should be surveyed prior to lifting operations to ensure that no obstructions can impede the positioning of the load.
- All lifting apparatus and associated equipment including crane wires, strops and lifting points should be checked prior to lifting and be certified appropriately.
- For subsea lifting operations the type of crane hook and crane hook latch has significant importance. The standard 'snap' latch hook can, because of their design, re-attach once disconnected, with the potential for a deployed load to be lifted after a positive disconnection. It is therefore necessary that for subsea lifting operations that the use of locked gate type hooks be used whereby the diver has to physically trigger the lock release.

- Lift bags utilised in association with subsea operations should be suitably certified and rated, with the capacity of the lift bags equal to, but not greater than the load. They should be clearly marked, provided with suitable certified and rated slings appropriate to the size of lift bag, have operable and working vent valves and inversion and operating lines (reference is made to IMCA D016).
- For lift bag operations, a dead man anchor may be attached to the load prior to commencement of inflation and a restraining line should be in place which will allow the lift bag to invert and empty in the event of an over inflation. In circumstances where the weight of the load is unknown, the buoyancy (lift bag inflation) should be conducted in incremental steps in order to control the ascent of the load.
- The use of divers in conjunction with lifting operations shall be kept to a minimum and ROVs should be used for as much as possible. However, in circumstances where diver intervention is necessary, they should maintain a safe distance from the load, associated rigging and lift bags, if used.

Hyperbaric Evacuation

The evacuation of the divers from a Dive Support Vessel can be divided into three distinct phases:
- Embarkation to and launching of the Self-Propelled Hyperbaric Lifeboat
- Immediate Assistance or Transit of the Self-Propelled Hyperbaric Lifeboat to a Safe Haven
- Arrival at a Safe Haven

Embarkation to and Launching of the Self-Propelled Hyperbaric Lifeboat

The chamber within the hyperbaric lifeboat is maintained at a pressure commensurate with the maximum depth of the living chambers with the trunking connecting the living chambers and the hyperbaric lifeboat chamber maintained at a lower pressure. This arrangement ensures that any loss of pressure within the hyperbaric lifeboat chamber will not result in the decompression of the living chambers.

The evacuation of the divers from the living chambers to the hyperbaric lifeboat chamber will only be carried out on agreement between the Master, Offshore Manager and Diving Supervisor with due regard to the safety of the divers. The primary and safest location for the divers is onboard the Dive Support Vessel in the living chambers and the decision to evacuate must be carefully considered.

On transfer into the evacuation chamber, the divers should be prepared for launch, and must be seated, strapped in and don safety helmets. The support crew onboard the lifeboat should consist of a marine coxswain for control and manoeuvring of the lifeboat and at least one Life Support Technician to maintain the diver's living environment during evacuation and rescue operations.

External to the hyperbaric lifeboat, the evacuation trunking will be isolated from the lifeboat chamber and marine crew will launch the lifeboat in co-operation with the marine coxswain, following pre-agreed launch procedures.

The Master of the Dive Support Vessel must, on launching of the hyperbaric lifeboat, ensure that a nominated support vessel (possibility another DSV operating in the vicinity or platform) is advised of the situation in order to ensure that sufficient support is available to the hyperbaric lifeboat once launched. At the planning stages of any diving operation, the contingency arrangements for the potential launch of the hyperbaric lifeboat should determine the location of an emergency umbilical or life support package, nearest safe haven and hyperbaric centres, support vessels and the location and limitations of a recovery lift beam for the lifeboat. These considerations will determine the extent of phase two of any evacuation operation.

Immediate Assistance or Transit of the Self-Propelled Hyperbaric Lifeboat to a Safe Haven

Once launched, the main objective will be to transfer the hyperbaric lifeboat and occupants to a safe haven at the soonest opportunity. Within the lifeboat chamber the divers environment will quickly deteriorate and the sooner they can be transferred, the higher their chances of survival.

Information

Life Support Package

A life support package is a self contained unit designed to be transported to a location to provide the necessary equipment and supplies to maintain a hyperbaric lifeboat throughout a decompression of the divers. This package may be stored on a support vessel, platform or shore location, dependent on the facilities available in the geographical area of operation. The package should include equipment necessary to provide life support functions to the divers within the hyperbaric lifeboat chamber until they can be transferred to a decompression chamber, such as electrical connections, gas lines, exhaust lines and communications.

After manoeuvring the hyperbaric lifeboat away from the Dive Support Vessel, the marine coxswain should be aware of the primary support location and the rescue options that are available. Such options may consist of a Dive Support Vessel operating in the vicinity, a platform or a shore based safe haven or hyperbaric centre. He should also be aware of the location of any life support packages and lifting beams to allow recovery of the hyperbaric lifeboat.

Dive Support Vessel Assistance

If a Dive Support Vessel is available in the vicinity, then immediate assistance may be possible. The availability of a Dive Support Vessel may be the preferred option, as the vessel will have the gas, equipment and facilities to provide life support functions or the capability to connect the hyperbaric lifeboat to the onboard living chamber complex. In addition, a Dive Support Vessel will have qualified, trained and experienced personnel who are familiar with diving operations and the care of divers.

Capabilities and lifting limits should be calculated for the particular hyperbaric lifeboat, lifting beam and offshore cranes (vessel and platform) to determine the feasibility of lifting the lifeboat from the sea onto the rescue vessel. If a lift is not feasible, the Dive Support Vessel may still assist, by providing additional personnel, equipment (such as additional radios), by protecting the lifeboat by means of providing a lee and chaperoning the vessel to a safe haven or by towing the lifeboat.

If the lifeboat can be lifted onto the Dive Support Vessel deck, the lifeboat may be connected to the chamber complex or life support package, if available.

Support Vessel or Offshore Installation Assistance

If a Dive Support Vessel is not available, the option to utilise another support vessel or offshore installation for assistance may be considered. Although specific equipment, systems and personnel may not be available, there may be the option of removing the hyperbaric lifeboat and its occupants from the full force of the environment.

As for a Dive Support Vessel, there may be the capability to recover the lifeboat, dependent on the availability of the lifting beam and the characteristics of the vessel or platform crane. Due to the height of platform cranes above the sea surface, it is unlikely that such cranes would be able to be utilised.

The availability of the life support package and suitable personnel to assist is a fundamental consideration. Life support personnel may be available from other rescued lifeboats or may be mobilised from a shore location.

If this option is not available, a support vessel may be able to provide a lee or tow the lifeboat.

No Assistance Available

If no immediate assistance is available, the hyperbaric lifeboat should be manoeuvred and maintained at the emergency location with any other survival craft. On arrival of rescue vessels, it may be necessary for the hyperbaric lifeboat to transit to a safe haven escorted by one of the rescue vessels.

Arrival at a Safe Haven

A safe haven can be defined as a location where there is perceived to be no further risk to the hyperbaric lifeboat, divers or crew. Such a location will be able to provide life support services and facilities to complete a controlled decompression. The most likely safe haven will be an onshore based location with hyperbaric chambers that can accommodate the transfer of the divers from the hyperbaric lifeboat, but may also be an offshore installation or vessel that provides such facilities.

Chapter 7

ROV SUPPORT VESSEL OPERATIONS

General Introduction

ROV operations can be considered to have similar hazards and operational limitations and constraints as diving operations. The only difference being that no human intervention is involved. However, this does not remove the fact that the ROV is a critical item of equipment and no damage or loss will be considered acceptable.

As all types of Offshore Support Vessel may have ROV capabilities and ROV intervention may be used during diving, pipe lay, survey or construction operations, there are general safety hazards, concerns and operational constraints that may be applicable:

- Code of Practice for the Safe and Efficient Operation of Remotely Operated Vehicles
- ROV Operational Planning
- ROV Operations and Environmental Conditions
- ROV Launch and Recovery
- ROV Operations in the Vicinity of Divers
- ROV Operations in the Vicinity of Offshore Installations
- ROV Operations in the Vicinity of Pipelines
- Emergency and Contingency Situations

Code of Practice for the Safe and Efficient Operation of Remotely Operated Vehicles

Due to the presence of human intervention, diving operations are closely regulated and very specific roles and responsibilities are placed on the diving contractor by the Health and Safety Executive. However, ROV operations are regulated by the industry in the form of Guidelines issued by the International Marine Contractors Association (IMCA). Central to these industry guidelines is the Code of Practice for the Safe and Efficient Operation of Remotely Operated Vehicles.

The purpose of this Code of Practice is to provide common standards, guidance and recommendations across the industry irrespective of the geographical location of the operations.

Figure 7.2 ROV Launch and Recovery

Figure 7.1
Kommandor Subsea 2000 – ROV Support Vessel

IMCA R004

Code of Practice for the Safe and Efficient Operation of Remotely Operated Vehicles

The Code of Practice contains guidance on ROV classification, tasks, tools, environmental considerations, operations, equipment certification and maintenance, personnel (qualifications and training) and responsibilities.

Reference

ROV Operational Planning

The planning of any ROV operation should consider not only the conditions and safety hazards at the proposed worksite, but the complete transit of the unit from the deck of the vessel to and from the site. The planning and risk assessment phase should therefore consider the vessel and sub surface environmental conditions, the capabilities of the ROV and ROV handling systems, the worksite conditions, obstructions and any foreseeable potential hazards and should consider emergency and contingency situations.

As summarised previously, ROV operations are varied and can include diver observation, installation inspection, cleaning and debris removal, pipeline inspection, seabed surveys, drilling support, subsea installation construction support, telecommunications support and location and recovery operations.

Each type of operation will therefore have specific planning considerations. However, the evaluation and assessment of any ROV operation should consider, at a minimum, the following:

- The type of ROV and the limitations and characteristics of the system, including the launch and recovery system.
- The position keeping capabilities of the vessel, taking into consideration the obstructions and hazards that may affect the ROV and ROV umbilicals during a vessel loss of position.
- The location of the ROV operation with due regard to any identified subsea obstructions or structures that may pose a potential hazard to the ROV and the main or excursion umbilicals.
- The environmental conditions on the surface and sub surface and the effects of these conditions on the vessel and the ROV.
- The vicinity of other vessels and, in particular, diving operations which may interfere or be interfered with by the ROV operations.
- Any simultaneous operations being conducted from the vessel including crane operations or DP operations (taut wires) which may impact the operation of the ROV. As with a diver's umbilical, the ROV main and excursion umbilicals are crucial components in maintaining the integrity of the systems.
- Umbilical management should, especially in deep water, consider the positioning of the vessel in relation to the ROV. The vessel should ideally be positioned upstream from the ROV to reduce the effects of umbilical drag and also assists in maintaining the umbilical clear of the vessel's structure and thrusters and propulsion systems and any potential to foul.

- Water intakes and discharges from offshore installations can cause turbulence and affect the ability to control the ROV when in close vicinity to such hazards. Discharges and intakes should be shut down during ROV operations.
- Pollutants of any kind within the sub surface environment can cause potential problems with regards to visibility and buoyancy (gas) of the ROV.
- Any operations being conducted on the surface installation such as pressure testing, crane operations or over side work which could result in the potential for dropped objects should be ceased.
- Communication between the ROV pilot and all other relevant parties such as bridge officers, dive control, crane operators and offshore installation should be maintained at all times. Permits must be in place when working in close proximity of installations to ensure that work is co-ordinated in a safe manner.

ROV Operations and Environmental Conditions

The current and forecasted external environmental conditions can have considerable implications to any ROV operation and can affect the ability to safely launch, operate and recover the ROV to the vessel. In respect of ROV operations, the environmental conditions are not restricted to the sea state on the surface, but must consider all of the following factors:

- Surface Sea State and Swell
- Surface Wind, Weather and Visibility
- Currents and Tidal Conditions (Surface and Sub Surface)
- Sub Surface Visibility
- Water Salinity and Temperature
- Water Depth

However, the effect that these environmental conditions will have on the ROV operations should not be considered in isolation. The combination of the individual factors may pose a particular hazard that when viewed in isolation may be regarded as acceptable and therefore a risk assessment considering all of the environmental factors is essential. These factors will be crucial in determining whether a particular ROV can be safely deployed in the prevailing conditions dependent on the power, manoeuvrability and type of ROV.

Surface Sea State and Swell

The purpose of the ROV Support Vessel is to provide a suitably stable platform from which offshore operations can be safely and efficiently conducted. The weather conditions at the surface can therefore reduce the stability of the platform which can pose a potential hazard to ROV operations. Adverse rolling, pitching or heaving of the vessel by the surface wave and swell conditions can, in particular, increase the risk for injury to personnel or damage to or overloading of equipment during launch and recovery operations.

Surface Wind, Weather and Visibility

Wind speed and direction relative to the aspect of the vessel may have considerable affect on the station keeping abilities of the vessel. In situations where the ROV operations are to be conducted alongside an offshore installation or structure, the ability to maintain station will be a major consideration. However, in any operation, the ability of the vessel to maintain position within predetermined limits is essential to avoid any excess or unexpected loads being placed on the main ROV umbilical or excursion umbilical.

Weather conditions at the surface such as rain, sleet, snow or fog may affect the ability of the vessel to remain on location close by an offshore installation with the effect that ROV operations will have to cease.

A combination or the surface weather conditions can, dependent on the protection provided to the ROV crew, result in a hazardous working environment on deck for the personnel involved in launch and recovery operations.

Currents and Tidal Conditions

Currents and tidal conditions can affect both the station keeping abilities of the vessel and the ROV whilst deployed. Similar to wind speed and direction, current and tidal conditions will affect the station keeping footprint of the vessel and therefore the ability to provide a stable platform for the launch, operation and recovery of the ROV.

During launch and recovery operations in strong current or tidal conditions there may be a risk of contact between the ROV and the vessel. The direction and strength of the current or tidal conditions should therefore consider the location of the ROV in relation to the vessel during launch and recovery.

For the operation of an ROV sub surface, the strength, direction and potential changes to the currents can sometimes be difficult to predict to a high degree of accuracy and will change dependent on the depth of operation. An ROV being deployed from the surface to the seabed may therefore be subjected to considerable changes in current strengths and directions.

Sub Surface Visibility

Good sub surface visibility is not only essential to allow the ROV pilot to perform the subsea inspection or intervention work that the vessel has been contracted to perform, but is also necessary to manoeuvre the ROV to and from the worksite. Although status information, including position location and depth of the ROV will be available to the ROV pilot, a visual check by means of the ROV mounted cameras will not be available.

Poor sub surface visibility will be most likely in shallow waters, areas of high tidal or current conditions or on the seabed where the ROV may be liable to disturb seabed sediment. The type of seabed will obviously also be a factor.

Water Salinity and Temperature

The variation in water salinity and density may vary quite considerably in tidal estuaries and in different geographical locations and may need to be considered if the vessel is mobilised to a new geographical location. Changes in the water density may affect the acoustic signals used for positioning information and may also affect the buoyancy and stability of the ROV. Extremes of temperature, in arctic or tropical regions for example, can adversely affect the ROV handling systems and the electronic components within an ROV unit.

Water Depth

An ROV will have defined water depth operational limitations which consider the water pressure at that operating depth in relation to the design and construction of the ROV. Operating an ROV at a considerable water depth will introduce potential hazards due to the increased main umbilical length and the associated drag associated with this increased length. In addition, navigational ranges for positioning beacons will be adversely affected.

ROV Launch and Recovery

The launch and recovery of an ROV can be considered the most hazardous parts of the entire operation. During these operations, the ROV will be in close proximity to the vessel and the main propulsion and thruster units. The potential for impact between the ROV and vessel will therefore be highest during deployment and recovery. In addition, the operation of the vessel's thrusters in close vicinity to the ROV can adversely affect the positional control that the pilot has on the system.

Although each and every ROV system will have specific procedures for the launch and subsequent recovery of the ROV unit, the following has been provided as a general overview highlighting some of the concerns, hazards and operational considerations during such an operation.

ROV Launch

All launch operations should be conducted with due regard to any operating procedures, pre-dive checklists and the vessel's permit to work system.

- Launch operations must be supervised by a suitably trained and competent ROV Supervisor.

- Prior to launch, a thorough risk assessment and ROV dive plan should be conducted taking into account the operational and environmental considerations.

- Communication between the ROV pilot and all other relevant parties such as bridge officers, dive control, crane operators and offshore installation should be maintained at all times. Permits must be in place when working in close proximity of installations to ensure that work is co-ordinated in a safe manner.

- Integrity checks should be performed on the ROV. These checks should ensure that no damage to the structure or equipment is evident, that all systems (such as manipulators, cameras and beacons) are secured and that all the required extra equipment for the proposed operation is fitted. ROV buoyancy should be negative so that in the case of an ROV failure or tether severance, the ROV will descend to the seabed and allow subsequent recovery.

- Power checks should be performed on the ROV. These checks should ensure that all the component systems, including the cameras, gyro units, manipulators, thrusters, lights and indicators are operational prior to the launch of the ROV. The gyro heading should be checked against the support vessel's heading.

- Integrity checks should be performed on the tether management system to ensure that there is no damage, the component parts are in good condition and secured. These checks should include integrity checks on the TMS frame, umbilical and umbilical terminations, ballast, lights, cameras and connectors, as fitted.

- Power checks should be performed on the TMS unit. These checks should ensure that power is provided to all component parts including hydraulic systems, control units, lights, latches and winch units.

- Winch integrity checks should be performed to check oil levels, counter settings, brake operation, umbilical condition, main lift wire, emergency stops and all functions. The mode of operation (constant tension or heave compensation for example), should be agreed prior to commencement of operations.

- Emergency location equipment and survey / position sensors should be fitted and operational.

- The water depth at the launch location should be confirmed prior to launch and a launch target depth agreed.

- All ROV crew involved in the launch operation should be fully briefed on the procedure and be provided with appropriate personal protective equipment, including suitable life vests.

- Prior to commencement of winch operations or the opening of the moonpool or hanger side door, permission to dive from the bridge crew must be confirmed. The bridge crew should also advise of any operations that are simultaneously being conducted or planned to be conducted during the anticipated duration of the ROV dive.

- For cursor handling systems, the ROV and cursor will be deployed with the weight of the system on the cursor until the cursor is fully deployed. The tension will then be transferred to the umbilical winch. The launch can then continue to the launch target depth.

- For A-Frame launch handling systems, the A-Frame will be extended fully clear of the support vessel prior to commencement of paying out the umbilical or main lift winch.

Figure 7.3   ROV Integrity Checks

ROV Recovery

All recovery operations should be conducted with due regard to any operating procedures and the vessel's permit to work system.

- Recovery operations must be supervised by a suitably trained and competent ROV Supervisor.

- The risk assessment and ROV dive plan should include recovery and emergency recovery operations and as such should take into account the operational and environmental considerations.

- Communication between the ROV pilot and all other relevant parties such as bridge officers, dive control, crane operators and offshore installation should be maintained at all times.

- All ROV crew involved in the recovery operation should be fully briefed on the procedure and be provided with appropriate personal protective equipment, including suitable life vests.
- The bridge should be informed prior to commencement of any recovery operation and any simultaneous operations that may interfere with the recovery should be suspended or reviewed.
- For recovery, the ROV should be maintained in auto heading mode with propulsion utilised to thrust fully downwards.
- For a cursor system, the ROV should be recovered to the cursor and then to the hanger, via the moonpool or hangar door, as applicable. The moonpool or hanger door can be closed and the ROV lowered to deck and secured. The cursor should be recovered to the stowed position and locked in place and ROV secured.
- For an A-Frame launched system, the ROV and TMS should be recovered with the A-Frame in the fully extended position. With the ROV and TMS recovered fully and stable, the A-Frame should be swung inboard in a controlled manner.
- If possible, vehicles should be washed with fresh water after a dive.

ROV Operations in the Vicinity of Divers

Simultaneous diving and ROV operations are a regular occurrence and, as with any task where human intervention subsea is required, can pose many hazards to the diver. ROV units may be utilised to assist the divers, with specific skid mounted tools, to visually monitor the divers or to survey the worksite for potential hazards and obstructions.

Close liaison between the dive supervisors, ROV pilots and bridge crew is essential in order to ensure that the particular hazards associated with these simultaneous operations are suitably addressed.

- The dive supervisor must always have authority over the ROV pilots whilst the dive bell or the divers are in the water. This authority should extend to recovery and deployment operations where permission to launch, move and recover the ROV must first be obtained from the dive supervisor and the bridge officers.
- Prior to commencement of operations, clear operating procedures including emergency procedures must be agreed between the dive supervisor and ROV pilots. All the potential hazards should be discussed at the risk assessment and planning stage. This assessment should include the competence and experience of the ROV pilots with regard to diving operations.
- Operations which require the deployment of saturation divers from a diving bell and ROV assistance have a high potential for entanglement of the diving bell umbilical, diver's excursion umbilical and the ROV tether and excursion umbilical. The umbilical lengths deployed should be closely restricted and monitored and guards should be fitted to the ROV thruster units to prevent damage to diver umbilicals or contact with the divers.
- Loss of position of the support vessel can result in the uncontrolled movement of the diving bell, divers and ROV units. The risk assessment and dive planning process should therefore consider the launch positions of each unit and their deployment and recovery paths to and from the worksite. The potential for the crossing of umbilicals and obstruction of the divers during such a loss of position and drift off should be considered and the deployment locations of each system should take this into consideration.
- Physical contact between the ROV unit and the divers can have a high potential for injury due to the power, weight and force (when moving) of the ROV. The close monitoring of the position of the ROVs and divers and close liaison between the control stations onboard the support vessel are critical to avoid such impacts.
- If the ROV has a TMS garage, it is unlikely that the garage will be visually sited during all operations. It is therefore important that its position is established and monitored at all times, especially in areas where visibility may be poor or where strong currents and tidal conditions may affect the ROV position. Any problems associated with the position of the TMS garage or ROV unit should be immediately communicated to the dive supervisor.
- There should be a direct communications link between the diving supervisor and the ROV supervisor (or pilot). In addition, the diving supervisor should be supplied with a repeater monitor showing the same picture seen by the ROV pilot.

Reference

AODC 32
Remotely Operated Vehicle Intervention During Diving Operations
AODC 32 provides guidance on the safety considerations that should be taken into account where simultaneous diving and ROV operations are being conducted.

- Risk of electrical shock to the divers should be fully considered and the guidance detailed in Code of Practice for the Safe Use of Electricity Underwater (AODC 035) should be followed.

ROV Operations in the Vicinity of Offshore Installations

For the purposes of this section, ROV operations in the vicinity of an offshore installation are considered to include operations directly concerned with the structure such as subsea structure inspection work, debris removal and seabed surveys.

- Irrespective of the work to be carried out, all subsea operations within the 500 metre zone of an installation must be conducted in accordance with the installation's permit to work and permission to commence an ROV dive must be obtained from the Offshore Installation Manager (OIM). This will ensure that any operations being performed on the offshore installation that may interfere with the ROV dive are ceased for the duration of the dive.
- The positioning of the Offshore Support Vessel within an offshore installation 500 metre zone is critical and must be thoroughly risk assessed taking into account the prevailing and forecast weather conditions. The position should not result in a potential drift on or blow on situation should the vessel lose manoeuvring propulsion.
- When in close proximity to an offshore installation, water intakes, discharges and pollutants, as previously highlighted, will be a particular safety hazard which should be assessed and discussed with the OIM.

General visual inspections are conducted to identify any structural damage or defects, to examine the condition of equipment and systems (such as anodes), to identify any debris or obstructions and to identify and examine any other areas of concern such as marine growth and corrosion.

Inspection operations will be recorded with a verbal record provided by the ROV pilot of any salient points.

- The position of any damage or deformity and an estimate of the extent (physical size, including depth) and a general description of the damage or deformity should be noted.
- The position of any debris should be noted. The type of debris and whether this debris has fallen from the platform may indicate other areas to inspect where the debris may have impacted. Excessive areas of debris or large items of debris may provide a potential hazard to the installation structure and therefore any details relating to the size of the debris field or individual items should be ascertained.
- The position and extent of any corrosion should be noted. If the ROV is fitted with suitable equipment, cathodic potential readings should be taken.
- The condition and percentage of wastage of any anodes and their attachment points on the structure should be noted. Any missing anodes or marine growth in the immediate vicinity of the anodes should also be recorded. Marine growth around the anodes may indicate that the anode is no longer effective.
- The position and extent of any marine growth should be estimated.
- Inspections of the structures may extend to seabed surveys around the structure and for any future installations and pipeline routes. These inspections are used to determine if there are any obstructions, debris or seabed scour.

ROV Operations in the Vicinity of Pipelines

ROVs may be utilised in the vicinity of pipelines for a number of purposes.

- Prior to a pipeline installation, the seabed and / or trench will be inspected to identify any obstructions prior to the pipeline installation operation along the pipeline route.
- During pipeline installation operations, the ROV may be utilised to assist the Pipe Lay Vessel by providing a visual display of the lay operation, to assist with the positioning of the pipeline heads into specific target areas, to ensure the pipe is laid within the pipeline corridor and to check for any potential damage during the installation.
- Subsequent to a pipeline installation, an as laid survey will be conducted noting the actual pipeline route and the position of any supports, sand bags, mattresses, joints, flanges, bends, valves and other pertinent information.
- Subsequent to a pipeline installation, the pipeline will be inspected at regular intervals in order to confirm the stability of the pipeline, identify any damage, debris or areas of corrosion. Pipeline damage can include coating damage, cracking, grooving and deformation. In addition, the condition of any protection devices should be examined.
- For pipelines which have been buried, any areas which have been missed during the burial or rock dumping process should be identified. A pipe tracker may be utilised to permit the ROV to follow the route of the buried pipe.

Operations in the vicinity of pipelines may require the Offshore Support Vessel to be positioned in close proximity to an offshore installation. However, the majority of such operations can be expected to be conducted in clear water locations. Irrespective of the vessel location, the permission to dive in close proximity to a pipeline will still require the permission of the appropriate OIM and be conducted in accordance with the offshore installation permit to work system.

- For operations where the ROV is required to follow a pipeline route, the ROV may be fitted with a wheeled skid or will free fly.
- For any pipeline survey, the ROV will generally be operated positively buoyant. When using the wheeled trolley sufficient down thrust is applied to keep the wheels in constant contact with the pipe.
- Deployment of the ROV to the worksite for pipeline operations should be planned with due regard to the environmental conditions and their effects on the vessel and ROV throughout the planned duration of the task as the vessel will be required to maintain position relative to the ROV throughout.
- The vessel should be positioned such that during the deployment operation, the ROV can be provided with a lee.
- If possible, the ROV should be deployed downstream of the prevailing current and at right angles to the pipeline, trench or proposed pipeline direction. This aspect will ensure that the ROV, in the case of a failure, will drift clear of the pipeline, rather than becoming a potential source of damage. The orientation of the ROV into the current will require a forward thrust to transit the ROV to the pipeline.
- With the ROV at the pipeline, the vessel can be positioned with due regard to the prevailing environmental conditions for optimum station keeping. With the ROV and support vessel stable, the pipeline operations can commence. In most cases, the DP system 'follow sub' mode can be utilised where the DP system can accept the ROV transponder as the only position reference system in order to maintain position relative to the ROV. As the ROV moves along the pipeline, the pre-defined operational radius centred on the ROV the vessel will be moved accordingly to maintain the radius, with the vessel following the ROV transponder.

Chapter 8

CONSTRUCTION OPERATIONS

General Introduction

Offshore construction operations have traditionally been conducted by specialised crane barges. However, most modern Offshore Support Vessels are now fitted with high capacity and capability heave compensated cranes which can fulfil a wide range of subsea and surface installation and decommissioning operations. The addition of such offshore cranes has resulted in vessels capable of multi-purpose operations without the need for additional vessel support.

The variety of structures that the Offshore Support Vessel may be required to install or de-commission can vary widely and therefore detailed analysis of the load, lifting equipment and environmental conditions will be required. The purpose of this section is not to describe this detailed analysis, but to provide general guidance on the marine aspects of any offshore crane and construction operations such as:

- Crane Modes and Functions
- Lift Planning
- Crane Wires

Figure 8.1   Offshore Crane Operations

Crane Modes and Functions

The offshore environment causes numerous hazards and challenges to crane operations particularly when divers may be involved. The control and stability of the load being deployed or recovered is paramount in order not only to protect any divers that may be in the water, but also to maintain the integrity of the load and to avoid any contact with structures or with the vessel itself.

Standard modes and crane functions fitted on modern cranes for construction operations include:

- Constant Tension Mode
- Passive Heave Compensation
- Active Heave Compensation
- Splash Zone Mode
- Intering Systems
- Gross Overload Protecting System (GOPS)
- Limits and Cut Offs

Constant Tension Mode

Constant Tension Mode is utilised to maintain a constant tension on the crane wire based on an input specified by the crane operator. This mode is most advantageous when a load has been deployed to the seabed. The crane operator can input the pre-determined set tension such that the load is not inadvertently lifted from the seabed.

Alternatively, constant tension mode can be used when there is a possibility that a seabed load may be in excess of what it should be. For example if a load is in a muddy seabed and suction may be a factor in the initial lifting of the load. However the mode should not generally be used for lifting loads off the seabed because there is the potential for any load to descend in an uncontrolled manner.

Heave Compensation

The aim of heave compensation systems is to maintain the load stable in relation to the seabed irrespective of

Reference | **Merchant Shipping and Fishing Vessels (Lifting Operations and Lifting Equipment) Regulations 2006**

The MS Regulations apply to all United Kingdom registered vessels and all vessels when operating within United Kingdom waters. The Regulations detail requirements for the use, maintenance, training, record keeping, testing, thorough examination and operation of lifting appliances.

the motion of the boom tip due to heave, pitch and roll. Passive and active heave compensation systems are common.

Passive Heave Compensation

A passive heave compensation system uses the mass of the load to reduce the motion of the load in relation to the seabed. Load balance can be achieved by a compensation cylinder which uses the motion of the load to charge accumulators which in turn recycle this stored energy to dampen the motion of the load.

This mode can be used to raise a load off the seabed. Activating a passive heave mode adds the gas buffer into the high pressure side of the main hoist hydraulic circuit. For example, if a load of 100 tonnes were to be lifted, but it was expected that an increase of 10 tonnes was likely due to mud suction and the gas buffer has a value of 15 tonnes, then the load could be lifted in this mode without any effect on the maximum load of the crane.

The use of this mode is likely to result in crane overload alarms. However, it should be noted that these overload alarms are set for the static SWL rather than the dynamic capability of the crane.

Active Heave Compensation

An active heave compensation system compensates for the effects of the vessel's heave, pitch and roll motions as calculated by a motion sensor in the boom tip or jib head. This involves real time automatic control of the winch or a compensation cylinder in order to counteract the heave motion of the jib head and such systems often make use of energy stored in nitrogen accumulators. The cylinder will balance and compensate the weight of the load during heave compensation.

The central processing unit uses these sensor inputs in combination with the actual slewing angle, boom angle and wire pay out status of the crane to make electric commands to the main hoist pumps to pay out or recover wire for stabilisation of the main hoist hook.

Note: Since an active heave compensation system slows down the hoist and lowering rate of the crane, the system should only be used when the load approaches the seabed or during initial recovery operations. Additionally, the use of the active heave compensation may lead to excessive wire wear.

Splash Zone Mode

This mode is utilized to lower or raise a load through the splash zone, or can be used to hold a load on the surface while water is drained from the structure. The splash zone mode results in a reduction in the relative movement between the load and the waves.

As with the passive heave mode, the use of this mode is likely to result in crane overload alarms. However, it should be noted that these overload alarms are set for the static SWL rather than the dynamic capability of the crane.

Intering Systems

Due to the heavy lifts that a Construction Vessel will be required to load and discharge in port and also deploy and recover at sea, there are significant stability considerations. With a heavy lift being deployed or recovered at the full extent of the cranes working radius, the vessel would heel significantly. In order to maintain the vessel in an upright position and thus minimise or negate the heeling effect, Construction Vessels may be fitted with anti-heeling systems. Such systems can consist of counter balance weights or intering ballasting systems.

Intering ballast systems can be either passive or active and are used not only for crane operations, but also for reducing the vessel's motions in a seaway.

In both passive and active intering systems a U-shaped tank is fitted and half filled with ballast water. Side tanks are connected by a water cross duct and air cross duct(s) are installed for air exchange between the side tanks. The dimensions of the tank will be defined such that the natural period of the tank (time for a full oscillation of the tank fluid) has about the same value as the shortest roll period to be expected in the service of the vessel. Stabiliser valves will be fitted to provide the active part of the system.

With a passive system, as the ship rolls or heels, caused by the external effects of the seas or movement of a heavy lift, the water in the intering tank will move to the counteract the effects of the roll or heel. The roll of the vessel is used to cause an oscillatory athwartships movement of the tank water. Thus the sea, in causing the vessel's roll, delivers the necessary energy to reduce the roll.

In an active system, the addition of the stabiliser valves provides a level of automatic control. Cyclic blocking of the tank water on the upwards moving ship side, by manipulation of the stabiliser valves, adapts the tank period to the actual ships period. The movement of the tank water is permanently tuned to counteract and reduce the roll.

Gross Overload Protecting System (GOPS)

In order to protect the winches and crane structure of the crane against severe overloads, Gross Overload Protecting Systems (GOPS) are often installed. In instances where the actual load were to exceed the safe working load of the crane by more than a set value, the

GOPS will release the applicable hoist winch against a hydraulic back pressure, thereby ensuring that a severe overload does not cause damage.

Limits and Cut Offs

On the winch drums for the main and auxiliary crane wires, software limits will be set for the end termination of the crane wire. This limit will include an allowance for a minimum amount of turns on the drum to avoid the wire being completely un-spooled.

At the head sheave or boom tip, a safety cut off limit will be installed to prevent the hook block being recovered into a position where contact with the head sheave or boom tip would cause damage to any part of the system. A simple example of such a system is an electro-mechanical limit switch suspended in chains (chandelier) below the boom tip sheaves. When the wire end socket reaches the chandelier, the hoisting is stopped by the control system.

Most cranes will also have a system whereby the crane hoist speed will be reduced when the crane hook is a pre-determined distance from the head sheave or boom tip. Recovery of the hook will initially be slowed once this distance is attained and then will stop when the wire end socket contacts with the chandelier.

The load chart for the crane should be programmed into the crane control system, and a maximum lifting capacity will be set. This should prevent the potential to overload the crane and is of particular importance on knuckle boom cranes where the load chart is constantly changing as the crane boom is luffed, slewed and extended or retracted.

Lift Planning

The detailed planning and preparation for a lifting operation should consider the environment (offshore/in port), the capabilities and capacities of the lifting appliance, the structure to be lifted, the lift path and the hazards and potential risks associated with such a lift.

The evaluation and risk assessment of the lifting operation should consider, at a minimum, the following:

- The risk assessment should be conducted and attended by all relevant personnel including the crane operator(s), deck supervisor, riggers and other relevant personnel, such as dive supervisors when simultaneous operations are planned.
- The status of the Offshore Support Vessels manoeuvring and positioning systems should be ascertained and any limitations discussed.
- The position of the Offshore Support Vessel in relation to any subsea or surface obstructions or structures may impact the lift path and limit the positioning of the crane and load. A review of the location and set up position for the vessel should therefore be completed before operations commence.
- Communication lines should be agreed and established prior to commencement of lifting operations. All relevant parties should have direct communications links including the crane operator, banksman, bridge watchkeeper, dive supervisor and ROV supervisor, if applicable. It is good practice for all lifts and essential for lifts where divers are deployed that a secondary communication system be in place in case of communications failure.
- Crane operators should be suitably qualified and competent for the particular crane being utilised.
- The capabilities, limitations and capacities of the lifting appliance should be considered at all stages of the lift. The safe working load of each component should be considered and the capability of the crane checked to ensure that the radius required allows the weight to be lifted without overloading the crane. Lifts should not be attempted that are out with the crane's lifting radius. Allowances for hydrodynamic loads and buoyancy should be made.
- All components of the lifting appliance and the lift rigging to be used for the lift should be certified and should be inspected prior to commencement of operations. All equipment should be used for the purpose for which it was designed and any item of equipment that is not certified should not be used and should be removed from service immediately.
- The design, weight and dimensions of the structure to be lifted should be considered with any particular snagging hazards identified. Lifts accepted for load

Information

Crane Failure Modes Effect and Analysis (FMEA)

The main aim of a crane FMEA is to identify and highlight the main components of the crane system (including electrical, hydraulic, and mechanical and wires) critical for the continued safe and efficient operation. This can be done by breaking down the various systems and then performing a criticality analysis based on the probability of failure (reliability), redundancy levels and the criticality (severity) of the failure. A comparison of the reliability, redundancy levels and severity would determine the acceptability of the failure.

out should be provided with a manifested total weight including the associated rigging. For subsea lifts performed offshore, an allowance should be made for the effects of seabed suction and for the added weight of any soil, such as mud sediment.
- The rigging weight should also take allowance for the crane wire. This is particularly relevant when a subsea lift is being deployed or lifted in deep water, where the weight of wire deployed may be significant.
- Rigging equipment should be designed for the largest dynamic hook load during the proposed lifting operation.
- Any instability of the load should be noted and assessed prior to lifting.
- The bridge must be aware of the details of the load (discharging or loading) including the weight, centre of gravity, final position and maximum radius of the crane during the lifting operation. This will allow the deck officers to calculate the stability of the vessel throughout the various stages of the lifting operation and determine the maximum heel during operations.
- The vessel should be as close to even keel and upright before the lifting operation commences and all tanks should be either full or empty to avoid the effects of free surface.
- The suitability of the deck position for placing of the lift if being loaded onboard should be checked. The strength loading including the position of any under deck stiffeners, beams and bulkheads should be determined and the lift placed accordingly to avoid localised concentration of the weight of the lift. Spreader beams may be used to distribute the weight.
- The lift path should be considered for all steps of the lift with any particular hazards assessed. At no time should the lift pass or be suspended above any personnel or subsea infrastructures (if at all possible).
- The proposed lay down position for the lift should be considered with due attention paid to any potential seabed obstructions or hazards. The potential impact of a dropped object situation during the lifting operation should be considered taking into account the size, shape, weight and anticipated drop zone for the lift if dropped.
- A dedicated banksman must be clearly identifiable for all lifting operations. The banksman must have a clear view of the lift and should, in circumstances where the lift may not always be visible to the crane operator, provide clear communication and guidance. The banksman should be in charge of the lifting operation unless divers or ROV units are in the water. In such circumstances, the control of the operation should pass to the dive supervisor or the ROV supervisor.
- The prevailing and forecast weather conditions including wind speed and direction and sea and swell state should be closely monitored. Limiting weather criteria should be clearly defined prior to commencement of operations.
- The watertight integrity of the vessel at all times during heavy lift operations is paramount to ensure the safety of the vessel and the crew. Watertight integrity checks should therefore be conducted prior to and during all lifting operations.
- High capacity offshore cranes are generally fitted with heeling compensation systems to maintain the vessel in an upright position during lifting operations. The status of any heeling compensation system should be confirmed prior to lifting operations.
- For lifting operations conducted whilst the Offshore Support Vessel is alongside a quay, suitable fendering should be in place and moorings should be continuously attended during the lift. Accommodation ladders may be lifted clear of the quay for particular lifts to avoid damage and the potential for injury to personnel as the vessel heel changes.

Crane Wires

For any offshore construction operation, the crane wire is critical and is a single point failure. Many systems on the crane will have redundancy, such as motors and pumps, but the loss of a crane wire can be considered a major failure.

The design of the crane and crane systems is essential in ensuring that the crane wire integrity is maintained once in use. However, the selection, installation, maintenance, inspection and thorough examination of the crane wire all require special consideration.

Winch Drum Groove Radius

For winch drums fitted with grooved surfaces, the diameter of the wire must be within specific parameters set by the drum manufacturer. The groove radius should be nominally larger than the wire rope. If the groove is too tight, then the wire will be liable to be deformed. Conversely, if the groove is too large (the wire too small for the groove), then there will be a lack of support for the wire.

Sheave and Main Winch Diameter

In addition to the grooves on the main winch drum surface, the diameter of the winch drum must also be considered in respect of the nominal wire diameter. This will ensure a tight stow when the initial wraps are spooled onto the drum.

Most crane wires will be reeved through a number of wire sheaves, particularly on knuckle boom cranes. The sheave diameter and the design and dimensions of the groove will affect the wear and fatigue on the wire rope. If the sheave groove is too narrow, the movement of the wire rope will be restricted, causing potential for damage. If the sheave groove is too wide, there will be a high surface pressure between the groove surface and the wire in contact.

The correct ratio of the diameter of the sheave to the diameter of the wire rope should be specified by the wire manufacturer and if followed, may increase the service life of the rope by up to 30%.

Wire ropes and sheave grooves are both inevitably subject to wear. As the diameter of a wire rope becomes smaller due to abrasion and stretch, it will wear out the groove to the smaller diameter of the worn rope. If a new rope is laid in such a worn groove, it will get wedged in the narrow groove and this will have a very adverse effect on its life.

Sheave Fleet Angle

The crane design should take into account the fleet angle of the wire during all crane operations and at all wire lengths. The fleet angle is the angle between the wire rope running to or from the extreme left or right of the drum and an imaginary line between the centre of the sheave, normal to the axis of the drum. If the fleet angle is too large, the wire rope will be pressed against the flange of the sheave. This will cause undue friction and wear on the sheave and the wire rope.

During spooling operations on smooth faced drums, an incorrect fleet angle will either result in the wire being spooled onto the drum with large gaps in the wrap or witll result in the wire bunching at one side of the drum. Subsequent wraps will be forced into any gaps on the previous layer or continue to bunch up.

Sheave Arrangements

The arrangement of the sheaves for the crane wire should be such that the wire should not be adversely affected by inappropriate bending during its life cycle.

The design should consider the number of bends. A single bend using a larger diameter sheave arrangement (figure 8.2) may be preferable to a double bend that would be present where two smaller sheaves are used (figure 8.3).

Continuous bending where the wire rope bends in one direction (figure 8.4) is preferable to reverse bending

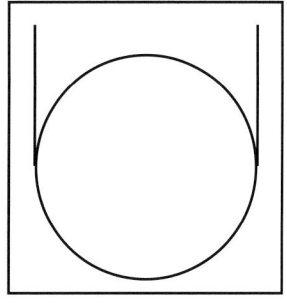 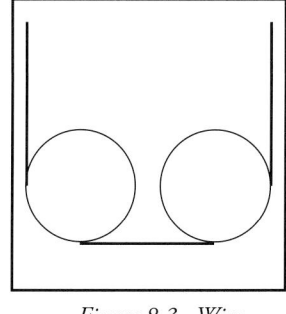

Figure 8.2 Wire Arrangement– Single Bend *Figure 8.3 Wire Arrangement – Double Bend*

where the wire rope is bent in the opposite direction to the wires natural rotation (figure 8.5).

If the distance between two sheaves is long compared to the dimensions of the sheaves (diameter) the steel wire rope gets the chance to rotate, following its origin (given during the production process) bending direction. If this distance is short, it is necessary to calculate with reverse bend, which means doubling the number of bends in this area.

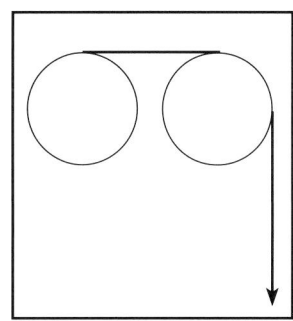 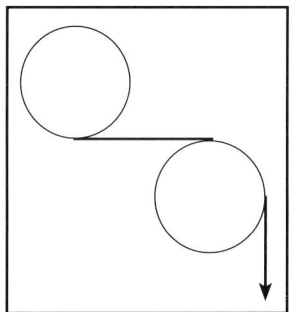

Figure 8.4 Wire Arrangement – Continuous Bend *Figure 8.5 Wire Arrangement – Reverse Bend*

Crane Wire Selection

A common mistake often made with regards to crane wire selection is to use the previous crane wire certificate as a starting point. If this method is used then errors can quickly occur, where a wire may be replaced with a slightly different nominal diameter or breaking strain. This may be acceptable given the parameters provided by the crane manufacturer; however, if a subsequent wire is purchased on the basis of the replacement wire certificate then the error will be compounded, leading to a wire diameter or breaking strain out with the acceptable limits advised by the crane manufacturer.

The following are the main criteria and factors that should be considered when selecting a crane wire rope:

- Nominal wire diameter taking into consideration any restrictions that the wire drum (lebus grooving), sheaves or blocks may require.
- Wire length required for all parts of the intended operation. It should be noted that the wire length on the drum will affect the load curves and therefore the crane lifting capacity. The need to consider all stages of the operation from loading in port to deployment at sea in the maximum water depth.

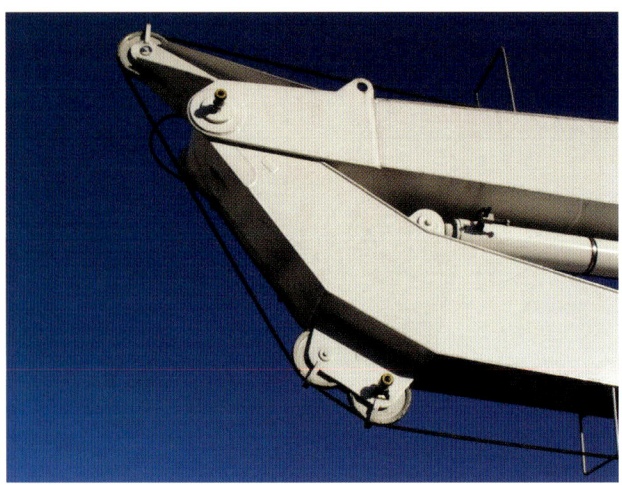

Figure 8.6 Knuckle Boom Crane – Sheave Arrangement

- The minimum breaking load, actual breaking load (as tested) and the safe working load of the wire should be considered.

- The purpose for which the wire rope is being purchased is a major factor. If the wire is to be used subsea, for example, then additional rope corrosion protection may be a requirement and the use of a non-rotating wire for deepwater operations.

- Wire end terminations will need to be compatible with the drum end connection point and more importantly, with the crane hook assembly to ensure ease of connection.

- Access to conduct maintenance such as lubrication of the wire may be restricted due to operational constraints and therefore the maintenance and lubricating requirements of the specified wire should be considered.

- Certification for the wire, sockets and all other ancillary components must be available and details of the minimum breaking load, actual breaking load, safe working load, nominal diameter and length must be available prior to installation to ensure that the correct wire has been provided.

Crane Wire Installation

The installation operation for a new crane wire can, if not managed correctly, have a high potential for damage to the wire and can pose a high risk to all those involved. It is essential that such installation operations are only conducted by competent personnel, familiar with the operation and equipment that will be used and within the scope of the vessels safety management systems (permit to work and risk assessment).

- All third party contractors required for the wire installation operation must be provided with a suitable safety induction and tour of the vessel. Such an induction must focus on areas of the vessel where the contractors will be working whilst onboard such as the after deck, crane winch spaces and other pertinent areas. Personal Protective Equipment (PPE) requirements in these specific areas must be indicated and the permit to work system and risk assessment processes explained.

- The entire operation including preparations, lifting operations, spooling, crane hook connection and load testing should all be conducted with due regard to the vessel's permit to work system and risk assessment. The risk assessment should be conducted with all relevent personnel in attendance including the sub-contractors, crane operators, deck crew and engineering crew responsible for the crane and crane systems.

- Communication systems should be in place prior to commencement of the installation with emergency signals clearly defined. A dedicated banksman to direct the crane operator and the spooling crew should be designated and clearly identified.

- The area where the operations are to be carried out should be suitably barriered off and all personnel made aware of the operation and the danger areas.

- Prior to the spooling device being loaded onboard the vessel, the lifting operation should be adequately assessed and a suitable lift plan prepared.

- The spooling machine should be secured on deck and positioned appropriately to provide a good lead from the spooling device to the crane winch drum with the direction of rotation of the wire being such that the wire follows its natural rotation. The fleet angle during spooling operations should be such that a tight, compact stow is possible and the introduction of reverse bends on the wire should be avoided.

- The specifications of the spooling device should be such that a back tension can be applied during spooling operations. The back tension to be applied should be discussed with the wire manufacturer prior to installation.

Reference

IMCA M171

Crane Specification Document

The construction of the wire will depend on its use. Ordinary lay or langs lay is used where both ends of the wire are secured (luffing wires for example). For single fall applications, then non-rotating wire rope should be used.

- Spooling on should commence with a forerunner being used to spool the wire end termination onto the drum. The wire end should be secured as per the wire drum manufacturer's instructions and spooling commenced at slow speed for the initial wraps in order to ensure a tight stow with no gaps.
- The spooling on operation should be conducted in a controlled manner with the back tension applied throughout the operation.
- Loose and uneven winding on a smooth drum should be avoided as such a situation can create excess wear, crushing and distortion of the rope.
- Bunching of the wire rope can lead to shock loads during crane operations, as turns drop into the gaps that bunching leaves in the stow.
- Slack turns can lead to kinks forming with the wire rope strands becoming permanently set in a distorted position.
- Over winding and cross winding on the drum should be avoided as this may lead to the wire rope digging in to proceeding wraps, causing abrasion and wire rope distortion.
- The calibration and re-setting of the limit switches and wire length counters and speed reduction functions should be completed at pertinent points during the spooling process. If practicable, these limits and functions should be tested and confirmed before the crane is deemed operational.
- The installation of the crane hook and assembly should be conducted in accordance with the crane hook manufacturer instructions and recommendations.
- On completion of the wire installation, it is advisable to run the wire to its working depth without a load and then repeat the process with a light load.

Figure 8.7 Incorrect Spooling can lead to Slack Turns

Chapter 9

PIPE LAY OPERATIONS

General Introduction

The details of the planning, methodology and execution of pipe lay operations is dependent on the equipment and systems fitted onboard the Pipe Lay Vessel. Each pipe lay system has its own particular idiosyncrasies, considerations and safety hazards. The methodology for vertical pipe lay operations will therefore vary quite considerably from a stern ramp reeled pipe operation, for example. However, many general considerations and concerns are common and it is those common factors which will be examined in some detail within this section. Particular reference will also be made to a stern ramp lay system in order to provide a case study in such a varied subject.

The following areas of pipe lay operations will be examined:

- General Safety Precautions
- Loading & Spooling Operations
- Pipe Lay Analysis
- Pipe Lay Surveys
- Pipe Lay Deployment Operations

Figure 9.1   Pipe Lay Operations – Stern Ramp Lay System

General Safety Precautions

Safety precautions during the spooling or loading of product and deployment and recovery operations can be separated into three distinct, though obviously linked, categories.

- Safety precautions primarily concerned with the welfare of the personnel working onboard the vessel and those assisting onshore, when spooling product.
- Safety precautions primarily concerned with the stability and integrity of the vessel including the minimising of the potential for damage to the vessel and the vessel equipment.
- Safety precautions primarily concerned with the integrity of the product during all phases of the spooling, loading, and deployment and recovery operations.

Figure 9.2   Seven Oceans - Pipe Lay Vessel

Safety of Personnel

Irrespective of the type of lay system onboard, pipe lay operations will involve personnel working at height, working in close proximity to rotating and moving machinery and working close by product and winch wires under tension.

- Due to the potential for injury to any personnel in areas where pipe reeling operations may be being conducted, it is essential that all areas underneath reeling operations and in the immediate vicinity are barriered off. This also applies to the areas around the spool base for loading operations, including any welding stations. Signs should be posted restricting access to authorised personnel only.
- All personnel working within the barriered off areas must be aware of the dangers from dropped objects, moving equipment (such as ramp fleeting

in a lay ramp system) and equipment under tension. Particular care must be taken close to the deck reel or carousel due to the possibility of a pipe break during reeling.

- Any safety devices such as ramp fleeting warning beacons, should be working correctly prior to commencement of operations.
- Personal protective equipment (PPE) must be worn by all involved. In particular, appropriate equipment should be utilised when working at height. This may be applicable for loading to a deck reel, deck carousel or under deck carousel.
- Clear communications should be established between the vessel deck, bridge, pipe lay control centre and, for loading operations, the shore based supervisor. For offshore deployment and recovery operations communications with the ROV supervisor and pilots and any third party vessels or facilities must also be agreed. Emergency stop and back up communications should be agreed before commencement of operations.
- A risk assessment should be conducted prior to commencement of operations involving all relevant parties including bridge crew, deck crew, pipe lay supervisors and any third party (shore based or offshore facility) personnel. This risk assessment should be followed up by safety briefings and tool box talks prior to and during the pipe lay operations.
- Visual monitoring of the main operating areas should be undertaken and remote (CCTV) monitoring of these and any inaccessible areas including the spool base when loading product should be maintained.
- A pipe arrestor may be used to prevent loose pipe moving away from a deck mounted reel. Prior to commencement of operations, the position of the pipe arrestor and its clearance from the product should be checked.

Stability and Integrity of the Vessel

The loading of product onto a deck or under deck mounted reel or carousel will result in a significant weight increase which will not only affect the stability, but also the trim and draft of the vessel. It is therefore essential that the deck officers are provided with accurate details of the total weight to be loaded including any reel weight if appropriate. An indication of the expected rate of loading of the product will also be beneficial in order to allow the deck officers to plan the necessary ballasting or deballasting operations that will be required during spooding operations. The effect the product will have on the GM of the vessel during all stages of the operation should be calculated.

Conversely, during deployment operations offshore, the weight onboard the vessel will decrease significantly and therefore similar considerations will be required.

For offshore operations, the Pipe Lay Vessel will be expected to operate in DP mode. For such operations, the tension and weight of the product being deployed from the vessel will act as an external force on the vessel and should therefore be considered. Some DP systems allow the input of an external force which can represent the pipe lay operation and therefore it is essential that the deck officers are aware of the tension at which the product is being deployed.

Integrity of the Product

The purpose of any pipe lay operation is to deploy the product to the seabed, in the correct location and without any damage. The integrity of the product is therefore of paramount concern.

- The dimensions and characteristics of the product must be considered throughout the pipe lay operation, ensuring that the pipe lay system is configured correctly for the product (clamp sizes, inserts on the reel or carousel, bending radius).
- Appropriate rigging suitable for the transfer of the pipe end from the quayside to the reel or carousel should be used.
- The configuration of the product on the quayside prior to and during loading (spooling) operations must be controlled to prevent deformation (sag) of the product. Approach rollers are used to support the product during reeling operations.
- A suitable securing methodology for the pipe end on the reel or carousel should be used to ensure that no damage can be caused.
- The alignment of the product between the quayside and the reel or carousel and between the carousel, reel and the tensioners during deployment must be controlled. Product damage due to poor fleeting or alignment can occur where the operation is not controlled and monitored effectively.

Loading and Spooling Operations

In some circumstances the product will be loaded onboard having already been spooled onto a deck reel or carousel. In such cases, the loading operation should be conducted as for any other heavy lift operation ensuring that the integrity of the product is maintained at all times.

However, spooling operations from a shore location onto a permanent vessel reel or carousel has specific hazards, operational and safety considerations, in addition to the general safety hazards identified in the previous section.

Figure 9.3   Pipe Lay Spooling Operations

- The tensionsers within the pipe lay system should be set in accordance with the project procedures and pipe analysis.
- The operation of all components within the lay system should be confirmed.

Spooling Operations

A very basic spooling operation for a stern mounted lay ramp system is summarised below. Although this methodology will be different for each variety of lay system, it will give an indication of the concerns during spooling operations.

- The ramp will be aligned with the shore approach rollers.
- The first pipeline stalk with pull head and lead string will be transferred from the spool base storage area to the approach rollers.
- The abandonment and recovery wire end is transferred ashore and connected to the pipe end.
- The first stalk is pulled towards the vessel until the stalk end is inboard of the open tensionsers. The tensioners can then be closed and the squeeze pressure set and the roller box closed around the pipe.
- The abandonment and recovery winch wire can be disconnected.
- The tensioners can then be used to push the pipe from the ramp to the deck mounted reel with the vessel's crane being utilised to support the pipe as it is transferred from the stern ramp to the deck reel.
- When the pull head assembly is aligned with the deck reel tie-in arrangement, the pipe end will be manipulated towards the securing point and connected using the tuggers, winches and cranes.
- The tie-in arrangement cover plates will be fitted in place. This process may require the deck reel to be rotated to allow access.
- Spooling operations proper can now commence with fleeting, speed and tensions monitored throughout.
- During spooling operations, the position of the pipe arrestor should be monitored and the pipeline chocked as required. Choking is the practice of forcing a smooth transition of the pipe, from wrap to wrap and layer to layer.

Preparations for Spooling

- The mooring of the vessel must be such that the vessel's position and aspect to the quayside will not alter during the spooling operation. Due to the configuration of many lay systems, the Pipe Lay Vessel may be required to moor stern first to the quayside with forward anchors deployed or moorings connected to buoys or dolphins.
- The prevailing and predicted environmental conditions including tides and currents should be assessed and moorings tended accordingly throughout the operation.
- Stern moorings should be deployed with due consideration to the changes in the fleeting of the product during spooling operations.
- For stern mounted lay ramp systems the exit roller will be positioned and utilised to assist with the support of the product during spooling operations.
- Checks should be made to ensure that the rollers within the lay system are of the correct type and size for the product being loaded.
- The vessel's crane and deck tuggers should be available to assist with the spooling operations, to support the product if required and for assisting with pipe transfer operations.
- A visual inspection should be conducted to ensure that no obstructions can interfere with the fleeting operations and the transfer of the product from the quayside to the onboard storage reel or carousel.
- Reel, carousel and fleeting ramp transit or sea fastening devices should be removed.

Information

Pull Head / Connection Assembly

A pulling head complete with pad eye is used to attach the end of the lead string, allowing connection to the main reel securing point.

Lead String

A lead string is required to provide a connection between the reel and the subsequent pipeline stalk to be reeled. The lead string length of pipe will be welded to the lead end of the first pipe stalk to be reeled.

- Spooling operations will be slowed as the end of the last stalk approaches the end of the approach rollers.
- The pipe end will be supported and the pipe will be spooled until the last stalk is transferred to the ramp.

Figure 9.4
Pipe Lay Loading Operations – Securing the Pipe End

Figure 9.5
Pipe Lay Loading Operations – Deck Mounted Reel

Pipe Lay Analysis

Pipe Lay Analysis is performed to ensure the integrity of the product during all stages of a pipe laying from spooling, straightening trials (if appropriate), pipe line initiation, deployment and laydown to abandonment and recovery operations.

The main function of the pipe lay analysis is to:
- Calculate and verify the loads involved in the load out operation to ensure that the vessel can be loaded safely and to ensure that the deck officers are provided with accurate stability information.
- Calculate and verify the loads on the approach rollers during the load out operation to ensure that no damage to the product can be caused due to failure of the rollers.
- Calculate and verify the positions and spacing of the approach rollers and the transfer of the product to the vessels reel or carousel.
- Calculate and determine the angle of approach of the product during spooling operations and therefore, in the case of a stern mounted ramp lay system, the optimum angle of the ramp throughout the spooling operation.
- Calculate the maximum loads that will be acting on the lay system rollers during spooling and subsequent deployment and / or recovery operations.
- Determine the tensioner settings for the product for the spooling on and subsequent deployment and recovery operations, taking into consideration the water depth at the proposed locations.
- Determine the maximum bending radius and bending moments for the product for all phases of the pipe lay operation to ensure that such limits are not exceeded.
- Determine the maximum winch tension that can be used for recovery operations.

Pipe Lay Surveys

Prior to the arrival of the Pipe Lay Vessel at the offshore site, there are considerable preparations that must be completed in order to ensure that the pipe lay operation can be conducted safely and efficiently, without damage to the product.

In particular, surveys are required to be performed prior to commencement, during and on completion of pipe lay installation operations.

These surveys can be geophysical (to determine the profile of the seabed), geotechnical (to determine the nature of the seabed soil) and visual monitoring during the pipe lay deployment operation. Typical survey requirements will include:

- ***Route Survey***

  The route survey is conducted in order to identify any obstructions or debris along a number of potential proposed pipeline routes. In addition the route survey can provide valuable information in respect of the seabed profile and nature for future trenching operations.

- ***Pre-Lay Survey***

  The pre-lay survey is a more focused survey along

the proposed pipeline route which will have been determined following the route survey. The pre-lay survey is more detailed in nature and will include the use of sonar to detail the seabed profile and the precise positions of any debris or obstructions.

- *Pipe Lay Survey*

The pipe lay survey is conducted in tandem with the pipe lay deployment operation and is used in order to monitor the touchdown of the pipeline on the seabed. The monitoring operation will generally be conducted using an observation class ROV and will ensure that the pipeline is deployed within the pre-determined parameters set by the client whilst checking for any potential damage such as buckling, coating or anode damage. For the purposes of such pipe lay surveys, the ROV will be fitted with positioning equipment to provide a high degree of accuracy.

Systems such as the 3D-Tracker application provide a two dimensional and a three dimensional real time visual display of the ROV, seabed and vessel during the pipe lay operation.

- *As-Built Survey*

The as-built survey is conducted on completion of the pipe lay operation and is used to verify that the pipeline has been deployed within the parameters set by the client. The survey is also used to check for any areas of damage or leaks which may require attention before any trenching operations are conducted.

- *Post Trench Survey*

The post trench survey is conducted on completion of trenching operations and is used to highlight any areas of concern on the pipeline. In particular any areas where there may be the potential for buckling can be identified.

- *Post Burial Survey*

The post burial survey is used to verify that the burial (protection) operation has been completed satisfactorily. The survey will identify areas where further burial, such as by rock dumping, may be necessary.

Pipe Lay Deployment Operations

The methodology used for the deployment of rigid and flexible products will vary from vessel to vessel. The purpose of this section is not to describe each and every possible system and its use, but to provide a general introduction to pipe lay operations and the basic safety and operational precautions during such tasks. The following details are therefore generic in nature, but biased towards a stern reeling ramp system for flexible product.

Figure 9.6 Pipe Lay Deployment Operations

- Prior to commencement of the pipe lay operation, arrangements should be made to ensure that suitable monitoring equipment is provided onboard the vessel to allow for accurate positioning of the vessel (survey package) and for touch down monitoring of the pipe (ROV support).

- To ensure that the pipe is not damaged during deployment operations, the vessel's DP System may be provided with a cable tension monitoring system. This system is fed with cable tension values measured by load cell sensors incorporated within the tensioners. This system will provide alarms at pre-set maximum and minimum values of tension.

- The pipe lay analysis calculations will determine the tensions and loads during deployment.

- The seafastenings are removed from the ramp and reel or carousel.

- The ramp is jacked to the pre-determined angle to ensure that the exit angle for the product is correct as determined by the pipeline analysis.

- The roller box should be closed, other than when deploying pipeline end terminations.

- A dead man anchor (DMA), clump weight or diverless latch will be positioned on the seabed as the start point for the pipe lay initiation with an attachment cable used to transfer the start-up head on the pipe end to the seabed. The use of the dead man anchor or clump weight and cable ensures that the deployment tension is maintained in the pipe as it is lowered to the seabed. The decision to use either a diverless latch or a dead man anchor or clump weight will be determined dependent on the subsea infrastructure at the location.

- The Pipe Lay Vessel manoeuvres along the pipeline route as the pipe is reeled from the vessel.

- As the vessel moves along the pipeline route and the pipe is deployed, the straighteners ensure that the pipe does not buckle and that the tensioners

maintain the deployment tension.
- The lay down process at the end of the pipeline route is conducted by using the abandonment and recovery winches. A lay down head is attached to the pipe end and attached to the abandonment and recovery winch wire and is lowered to the seabed.
- A controlled abandonment procedure can be applied when a pipe lay operation is prematurely stopped due to severe weather conditions, mechanical failure or an emergency station situation arising that requires the pipeline to be lowered to the seabed immediately.

Figure 9.7 Skandi Navica - Pipe Lay Vessel

Chapter 10

DYNAMIC POSITIONING SYSTEMS

General Introduction

Dive Support, ROV, Construction and Pipe Lay Vessels all require a high degree of position accuracy and station keeping capabilities. The ability to maintain a pre-determined position and heading are essential for safe and efficient operations.

At its very basic concept, dynamic positioning is a system that is utilised to maintain a vessel in a designated position with a designated heading in order to provide a stable platform for diving, construction, ROV, or pipe lay operations.

Any vessel has six freedoms of movement, namely yaw, surge, sway, heave, pitch and roll. The heave, pitch and roll of the vessel cannot be controlled by a DP system, however it is the function of the DP system to automatically control the yaw, surge and sway and therefore maintains the vessel in the desired set-point location or maintains the required heading control.

Yaw

Yaw is when the vessel rotates about the vertical axis, resulting in a change in the vessels heading.

Sway

Sway is when the vessel moves in a linear lateral (side to side) motion.

Surge

Surge is when the vessel moves in a linear longitudinal (fore and aft) motion.

Heave

Heave is when the vessel moves in a linear vertical (up / down) motion.

Pitch

Pitch is when the vessel rotates about the transverse (athwartships) axis.

Roll

Roll is when the vessel rotates about the longitudinal (fore and aft) axis.

Dynamic Positioning Systems

A generic DP system, used to control the vessel's position and heading, consists of a central processor linked to position reference systems and environmental sensors. The central processor utilises and sets power settings on the vessel's propellers and thrusters in order to maintain the vessel at the desired set point position and on the required heading. The environmental sensors are utilised to incorporate environmental forces into this control system.

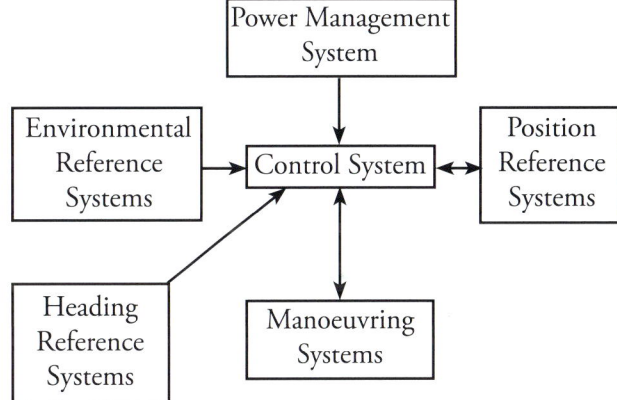

Figure 10.2 Typical Dynamic Positioning System

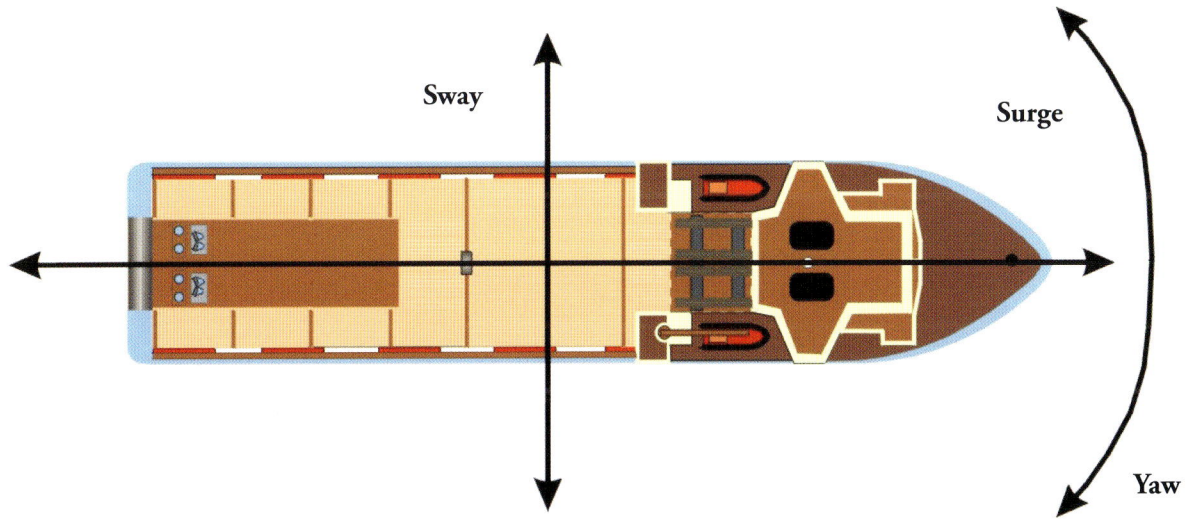

Figure 10.1 Forces involved in Dynamic Positioning

Basic System Elements

Generically, all Dynamic Positioning Systems consist of six elements:-
- Control System
- Position Reference Systems
- Heading Reference Systems
- Environmental Reference Systems
- Power Management System
- Manoeuvring Systems

Control System

The control system of the Dynamic Positioning System can be considered to consist of, not only the central-control system, but also the DP Operator and the Bridge Operating Consoles. The redundancy and configuration of the control systems will vary dependent on the vessels required operations. However, the basic requirement for the control system is to act as a central operating unit. The control system is provided with information from the reference systems, heading reference and environmental sensors and provides information to and receives information from the manoeuvring systems in order to maintain the vessel's desired position or heading.

Position Reference Systems

Position reference systems are required in order to provide the control system with a set point or location. There are currently five main types in use:
- Differential Global Positioning System (DGPS)
- Hydro-Acoustic Position Reference Systems
- Fanbeam
- Taut Wire Systems
- RadaScan

The advantages and disadvantages of the available position reference systems will depend on numerous factors such as the vessel type and configuration, area of operation and the availability of alternative reference systems. It is therefore very difficult to detail specific merits and drawbacks for each individual system; however the following section provides the very basic details and specifications for each position reference system type.

Differential Global Positioning System (DGPS)

The Global Positioning System (GPS) is the most common reference system used for Dynamic Positioning purposes and is to be found as a basic element of all DP systems. The system operates by means of reference satellites orbiting the earth in known positions, transmitting radio signals. These signals can be received by the GPS receivers fitted onboard. As the positions of the satellites are known, and the time period for the signal transfer can be calculated, the GPS receiver can determine the range or distance of the receiver from the satellite. If three satellite signals are received, then a latitude and longitude position can be calculated.

However, the use of the Global Positioning System for operations where criticality of position maintenance is paramount, such as diving, is not acceptable due to the level of accuracy (approx. 100 metres).

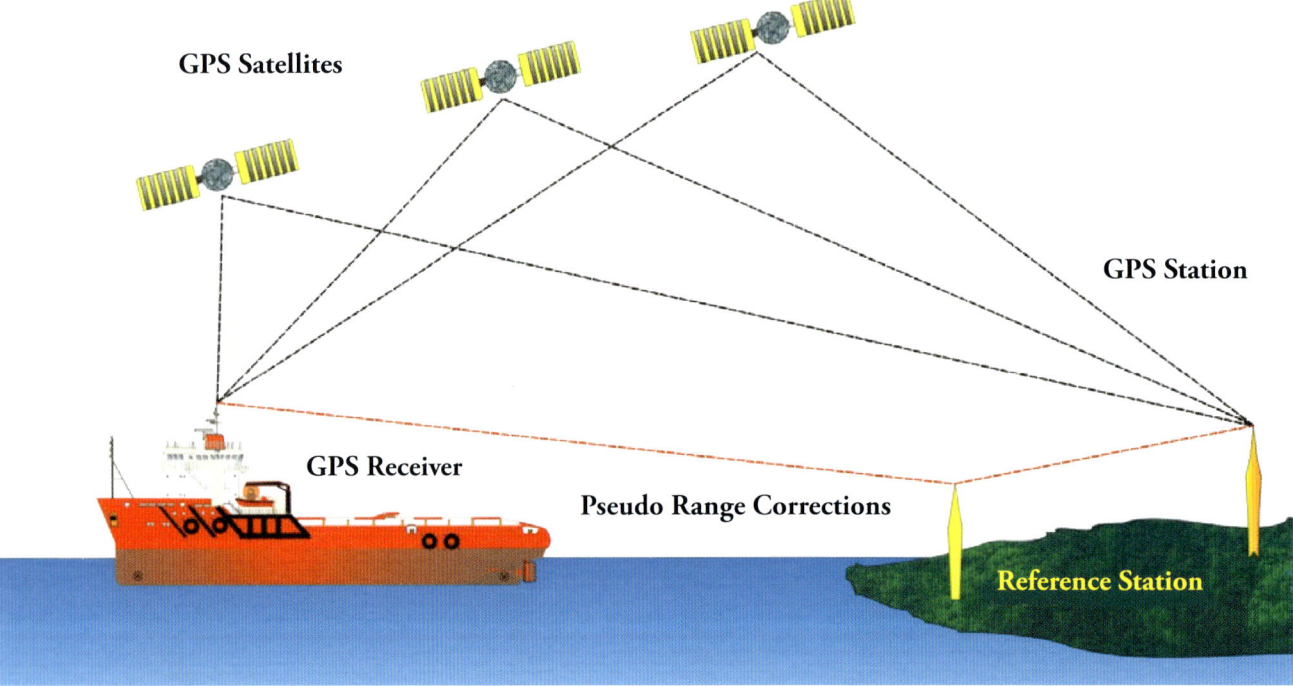

Figure 10.3 Differential Global Positioning System (DGPS)

To increase the accuracy of the position provided by GPS, a shore based reference station, at a fixed and accurately known location is established. As the location of the reference station is known, the GPS fix can be interrogated and a correction calculated, by the reference station, to improve the accuracy of the fix and thus be transmitted to the vessel. This correction is known as the 'differential correction' and hence the Differential Global Positioning System (DGPS).

DGPS has been proved to be most reliable in open water situations where there are no obstructions for the signal path. However, when alongside a structure, such as a platform or drilling rig, the line-of-sight satellite path may be interrupted causing signal degradation.

Hydro-Acoustic Position Reference Systems

Hydro-acoustic position reference systems use a vessel-mounted transducer (transmitter and receiver) and a transponder located on the seabed.

A pulsed signal from the vessel's transducer will activate a response signal from the seabed transponder, which when received by the vessel, will allow the transducer to calculate the position of the seabed transponder relative to the vessel's position. The position of the vessel is therefore known. HPR (Hydroacoustic Positioning Reference) and HiPAP (High Precision Acoustic Positioning) systems are standard on many DP vessels and are based on these principles.

The main problems associated with the use of these systems is that appropriate transponders require to be deployed and also that, as the systems are based on sound waves. Noise from the vessel can cause interference in water with differing temperature layers.

Fanbeam

Fanbeam utilises an 'optical laser radar' system that has a wide (20°) vertical beam. This allows the unit to track a fixed reflector target positioned on a stationary target, thus deducing the range and bearing from the fanbeam to the target reflector.

The fanbeam laser unit should be situated in a location that affords an unobstructed view of the target reflector location. The target reflector should be placed on a fixed unit in a position that, to the best possible degree, affords direct laser contact between the scanning laser unit and the reflector.

As the system operates using laser contact, range is limited to a maximum of some 2000 metres, however atmospheric conditions such as rain, sleet, for or snow will further reduce the capabilities of the system.

When using the system for position referencing, the following considerations should be taken into account:

- The positioning of the fanbeam laser unit and target reflectors is critical in order to ensure an unobstructed line-of-sight.
- The fanbeam laser unit should be positioned in an area where, as far as possible, it is not subject to

Figure 10.4   Transponder Beacons

Figure 10.5   Fanbeam Laser Unit System

excessive water spray or exhaust smoke. Similarly, heavy rain, snow or fog obscuring the line-of-sight can cause degradation of the system and, as detailed above, may reduce the range of the system.

- The fanbeam laser unit must be accurately aligned to the vessels fore and aft centreline in order to provide accurate relative bearings of the target reflectors.
- Degradation and loss of the system as a reference unit may occur when the vessel is pitching and / or rolling excessively.
- Bright lighting or reflective surfaces close to the target reflector may cause the loss of the target reflection or cause false target reflections.
- For more accurate heading control, alongside a moving installation such as an FPSO, multi-reference fanbeam may be required.

Taut Wire Systems

A standard taut wire system consists of a deck mounted davit arrangement. A depressor or clump weight arrangement on a wire fall is attached to the davit arrangement via a constant tension winch. With the wire deployed in the water and the clump weight resting on the seabed, angle sensors located on the davit arrangement boom detect the angle of the wire leading from the davit to the clump weight.

Once the clump weight is on the seabed, the constant tension winch will maintain the constant tension by recovering and deploying wire as necessary.

The length of wire available and the operating water depth will obviously determine the practicality of using the system for a particular operation.

Figure 10.6 shows a Kongsberg Simrad Mark 12 taut wire arrangement. The system comprises of a self tensioning winch attached to a seabed clump weight via a 5 mm steel wire rope. At the head of the boom is a gimballed sensor head, which measures the wires deviation from the vertical in the athwartships and fore and aft directions. The system has a maximum operating depth of 500 metres.

In using taut wire systems as a DP reference unit, the following considerations should be taken into account:

- Taut wires have operational limitations on the wire angle at the boom. This angle limitation is due to the fact that the more acute the angle from the boom end to the clump weight, the more likely it is that the clump weight will be dragged along the seabed. This may cause loss of the functionality of the system for DP purposes and possible damage to the arrangement.
- For deployment and recovery of the taut wire system, a clear unobstructed seabed is essential to

Figure 10.6 Taut Wire Arrangement

avoid damage to any subsea structures or to the taut wire itself. In addition, suction forces from seabed soils such as mud can cause particular problems.

- For deployment and recovery operations, a pendulum effect is possible on the clump weight. It is therefore advisable to lower the clump weight in a slow controlled manner until the clump weight is submerged. Similarly, when recovering the clump weight, the speed of the recovery operation should be slowed or even stopped, until the weight can be recovered without excessive rolling of the vessel and associated movement of the weight.
- Special consideration must always be given to other operations in the immediate vicinity that the taut wire system may interfere with, such as diving, ROV, crane or survey operations.
- A major problem associated with the actual operation of the taut wire system is the reduction in accuracy when utilised in areas with strong tidal forces. The taut wire will be susceptible to movement imposed

by any tidal streams and therefore inaccuracies will be present. This will be most apparent at slack water, as the tidal forces will be changing and at their most unpredictable.

RadaScan

The Radascan is a microwave based position sensor designed for close range operations between 10 and 1000 metres. Due to the microwave based system, the RadaScan system can overcome some of the operational limitations that can affect laser based systems, such as the Fanbeam or Cyscan.

The RadaScan system comprises a vessel mounted sensor, control and display unit and one or more retro-reflective transponders which are mounted on the target installation. The system provides accurate range and bearing measurements between the vessel based sensor and the transponder(s).

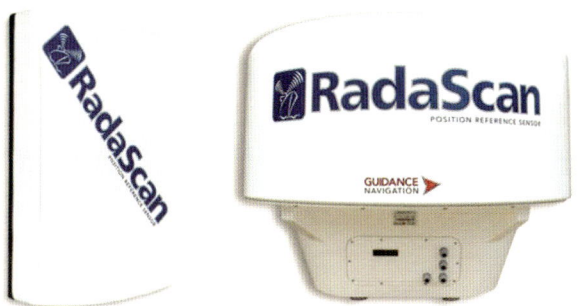

Figure 10.7
RadaScan Transponder

Figure 10.8
RadaScan Sensor

The system has a number of advantages, some of which are summarised below:

- Provided that the positioning of the sensor onboard the vessel and the transponders are unobstructed, and therefore no blind sectors are evident, the system

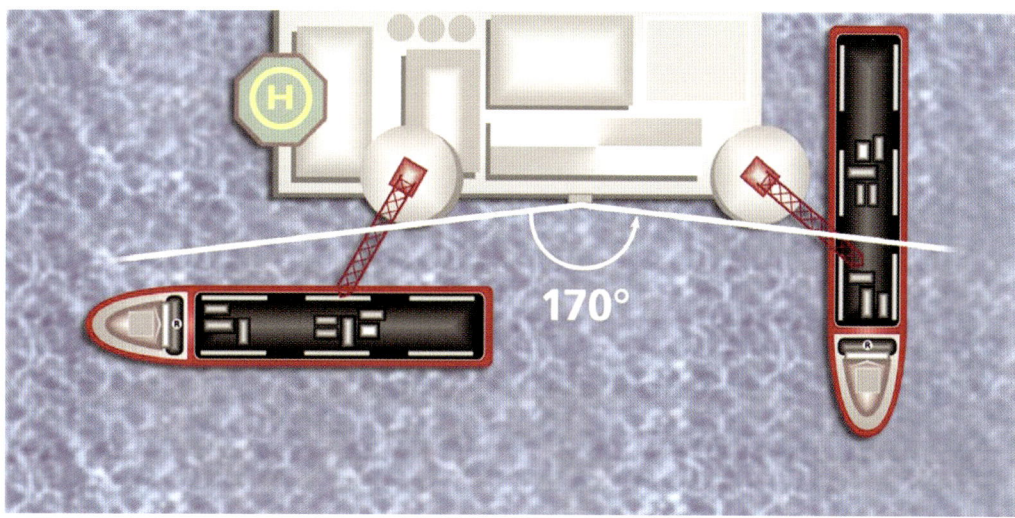

Figure 10.9   RadaScan 170º Diagram

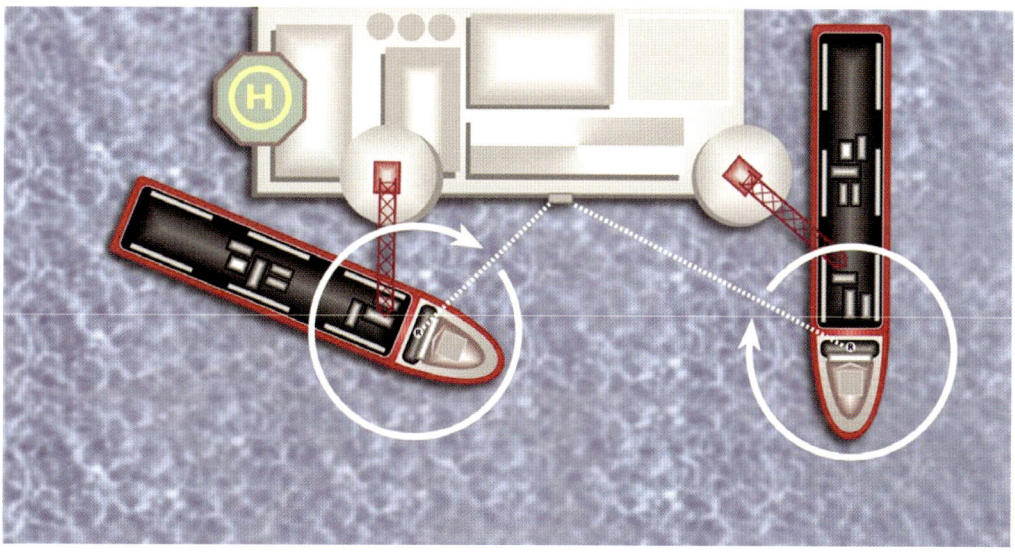

Figure 10.10   RadaScan 360º Diagram

OFFSHORE SUPPORT VESSELS   99

can operate over 360 degrees, minimising the need to move the unit dependent on the vessels set up alongside an installation.

- The system is not affected by environmental conditions which can cease fanbeam operations and the system can operate effectively even in heavy rain, snow or fog with no reduction in performance.
- Each system transponder is installed with a unique code. This coding ensures that positive identification of the transponder is provided, ensuring that there can be no ambiguity with regards to position information. In addition, the sensor will only reflect the signals from the coded transponder; therefore the system does not receive signals from false reflections or suffer from clutter as would be expected on a traditional radar system.
- The transponders are portable and self contained with their own internal battery pack with a life of circa two years and can be temporarily or permanently installed.
- The design of the transponders ensures a horizontal (azimuth) viewing angle of up to 170 degrees which provides excellent directional coverage for vessel manoeuvres and means that very few transponders will be needed to ensure full coverage at an installation.
- A single transponder can be used by more than one vessel simultaneously and up to five transponders can provide information to one vessel sensor simultaneously.

Heading Reference Systems

The heading reference for any DP system comprises of a number of gyro compass units. Generally two or three gyro units will be available to ensure adequate redundancy levels. The other vessel factors that will affect the ability of the DP System to maintain the vessel's position to the highest possible degree of accuracy are the vessel's pitch and roll. A vertical reference unit (VRU) is generally installed close-by the centre of rotation of the vessel to provide the pitch and roll data for the control system.

Environmental Reference Systems

The environment affects the performance of any DP System by adding an external force to the vessel's movement. In order to compensate sufficiently for this external force, wind sensors are provided which feed data to the DP System. This allows the control system to build up a model of the external wind force and therefore compensate for the anticipated movement, thus allowing the response times of the system to be reduced. This ensures that the accuracy of the position keeping of the vessel is maintained. The wind sensors must be located in an area free of obstruction and free of turbulence from ship's structures. For helicopter operations, the wind sensors should be de-activated to ensure that turbulence does not give a false wind reading to the DP System.

Power Management System (PMS)

The management of the power required to maintain a DP vessel and DP system operational is a complex subject. However, the basic function of such a power management system is to maintain power to the manoeuvring systems. Even in the event of a failure, the power management system should maintain power to the manoeuvring systems for as long as is possible.

For vessels that have equipment, other than DP related, such as deck cranes or A-Frames that require a high degree of power, the power management system ensures that load is shed and the manoeuvring systems prioritised to ensure the safety of the vessel.

This subject is covered in more detail in *Ship Dynamics for Mariners*, also published by The Nautical Institute.

Manoeuvring Systems

The vessel manoeuvring systems that can be utilised to provide the required control and movement for a DP system include the main propellers, rudders, tunnel thrusters and azimuth thrusters. The DP control system will utilise the directional qualities of the various elements and adjust the individual power settings in order to maintain position and heading. The configurations of the systems available are extensive and each individual system will have its own characteristics and limitations.

DP Classifications

The Classification of DP Systems is conducted by the relevant Classification Society in accordance with IMO Guidelines. In addition, there are also standards for DP Operations provided by National States, such as the Norwegian Maritime Directorate (NMD). For Det Norske Veritas (DNV) there are five classification groups for DP Vessels.

A summary of the DP Classes available and their relationship to each other is shown in figure 10.13:

DNV Classification		NMD Classification	
DNV notation	DP System Specifications	NMD Notation	NMD Class
T	A semi-automatic position keeping system without redundancy	DYNPOS T	Not Applicable
AUTS	An automatic position keeping system without redundancy	DYNPOS AUTS	NMD Class 0
AUT	An automatic position keeping system with a remote thrust control back-up and a position reference back-up	DYNPOS AUT	NMD Class 1
AUTR	An automatic position keeping system with redundancy in technical design	DYNPOS AURT	NMD Class 2
AUTRO	An automatic position keeping system with redundancy in technical design and physical arrangement	DYNPOS AUTRO	NMD Class 3

Figure 10.11 DP Classification (DNV and NMD)

Lloyds Notation	DP System Specifications
DP (CM)	This notation may be assigned when a unit is fitted with centralised remote manual controls for position keeping and with position reference system(s) and environmental sensor(s). It denotes that the machinery and control engineering equipment has been arranged, installed and tested in accordance with Lloyds Rules or equivalent.
DP (AM)	This notation may be assigned when a unit is fitted with automatic main and manual standby controls for position keeping and with position reference system(s) and environmental sensor(s). It denotes that the machinery and control engineering equipment has been arranged, installed and tested in accordance with Lloyds Rules or that it is equivalent.
DP (AA)	This notation may be assigned when a unit is fitted with automatic main and automatic standby controls for position keeping and with position reference system(s) and environmental sensor(s). It denotes that the machinery and control engineering equipment has been arranged, installed and tested in accordance with Lloyds Rules, or that it is equivalent.
DP (AAA)	This notation may be assigned when a unit is fitted with automatic main and automatic standby controls for position keeping, together with an additional/emergency automatic control unit located in a separate compartment and with position reference systems and environmental sensors. It denotes that the machinery and control engineering equipment has been arranged, installed and tested in accordance with Lloyds Rules or that it is equivalent thereto.

Figure 10.12 DP Classification (Lloyds Register)

IMO DP Class	NMD DP Class	Lloyds DP Class	DNV DP Class
—	—	DP (CM)	T
1	0	DP (AM)	AUTS
1	1	DP (AM)	AUT
2	2	DP (AA)	AUTR
3	3	DP (AAA)	AUTRO

Figure 10.13 DP Classification – Comparison Summary

Failure Modes and Effects Analysis

The DP System onboard any vessel should be designed and fitted in accordance with the requirements of the International Maritime Organization (IMO) and the relevant Classification Society. Furthermore, once installed, the DP System should be subject to a Failure Modes and Effects Analysis (FMEA) in order to:

- Identify all systems and sub systems and their mode of operation.
- Identify potential failure modes and their causes.
- Evaluate the effects on the systems and sub systems of each failure mode.
- Identify measures for eliminating or reducing the risks associated with each failure mode.
- Identify trials and testing necessary to prove the conclusions.
- Provide information to the operators and maintainers so that they understand the capabilities and limitations of the system to achieve best performance.

The FMEA is a requirement of most, if not all, Classification Societies as part of the acceptance criteria for IMO Class 2 and Class 3 DP Vessels and will therefore be required prior to the issue of the flag State Verification and Acceptance Document. However, the benefits to the vessel crew with regards to knowledge of the system they are using, and the confidence that an extensive FMEA will give to owners, managers and

IMCA M166

Guidance on Failure Modes and Effects Analyses

An FMEA should be conducted on completion of the DP System and should include practical tests and their results. An extensive proving trials program based on the FMEA should be performed on completion of the FMEA report. Further updates should be made whenever any equipment, system or sub system is modified.

Reference

charterers should be highlighted. It should be stressed that the FMEA should not be a 'one-off' document or report that is completed once the DP System has been installed and thereafter not re-visited. Any modifications to the vessel's manoeuvring capabilities, physical structure or DP components may affect the outcome of the FMEA and therefore the FMEA should be re-assessed following any such modifications or upgrades.

Flag State Verification and Acceptance Document

The International Maritime Organization (IMO) document 'Guidelines for Vessels with Dynamic Positioning Systems' was issued in order to provide an international standard for Dynamic Positioning Systems on all types of new vessels.

Compliance with the requirements and recommendations of the guidelines should be documented by a flag State Verification and Acceptance Document. This is a requirement for vessels built after 1st July 1994.

The overall objective of the flag State Verification Acceptance Document is to:

- Provide a comprehensive and safe testing and checking programme for the DP System.
- Demonstrate that the DP system is maintained to fulfil the requirements of the vessel's capability and integrity.
- Reduce over testing of systems such that components are not unnecessarily stressed.
- To ensure new problem areas are quickly incorporated into the system and resolved.
- To provide a continuous and structured record of events which are relevant to the DP operation.
- Provide a stand-alone document that can be easily audited by the flag state representative or by the organisation authorised by the flag state.

Proving Trials and Annual DP Trials

It is standard practice for proving trials to be conducted following or as part of the FMEA study. The proving trials are the practical testing and verification of the expected effects of failures to the DP System and form the basis for the subsequent annual DP trials.

Annual DP trials are based on the information obtained from the vessel's FMEA and on the previous annual trials. They are conducted in order to demonstrate that the vessel meets the requirements of the 1994 IMO 'Guidelines for Vessels with Dynamic Positioning'.

The trials constitute two parts:

- A review of the onboard DP documentation, including manuals, records of DP operations, maintenance and incidents, and the previous annual trials and FMEA.
- A series of practical failures in order to demonstrate and verify the DP Systems redundancy levels and failure effects for the power management system, propulsion and manoeuvring systems and the DP Control System, including all reference units.

The practical aspect of these failure trials not only provides verification of the systems onboard, but also provides the auditor with information regarding the level of knowledge of the vessel crew. In addition, the annual trials can be viewed as a valuable learning process for all personnel onboard who may be involved with DP operations.

It is standard practice for the annual trials to be conducted by an independent third party with representatives of the DP System manufacturer also present to offer advice and to perform any updates that may be required to the system software. Any such updates should be taken into consideration for the trials.

The proving trials and annual trials will generally be conducted under the following trial parameters:

- Whenever possible, all tests are to be carried out

127 DPVOA

Guidelines for the Issue of a Flag State Verification and Acceptance Document

The DPVOA guidance document sets out the procedures and requirements that a Flag State Authority expect, in order to issue a flag State Verification and Acceptance Document.

Reference

on full DP with some varying load on the system induced by movements of the vessel.

- During the failure tests, the system should not be retained until the DP operators; ECR staff and surveyor are satisfied that they understand the full effects of the failure and that all the information or indications to show what has occurred have been adequately recorded.

- When reinstating systems after failure simulations, two persons should check that circuit breakers, pump settings, automation etc. have been reset properly. A mistake in resetting or configuring the system could lead to failure effects worse than the design intent.

Personnel Training and Certification for DP Operations

With regards to Dynamic Positioning Systems, it is not only the DP Operators that are designated as 'key personnel'. As the DP system incorporates the engine room machinery and control systems, the vessel's engineers and electricians are also defined as 'key' and should also meet the minimum requirements for training, certification and experience as recommended by IMCA M177.

Training and Certification

The training of DP Operators is divided into four parts:

- Attendance and satisfactory completion of an approved induction course.
- Thirty (30) days sea-going DP familiarisation.
- Attendance and satisfactory completion of an approved simulator course.
- Satisfactory completion of six months DP Operations.

Additional requirements are to be fulfilled prior to issue of an 'unlimited' (i.e. Class 2 and Class 3) DP Operators Certificate. The Nautical Institute developed the training system accredit over 40 DP training centres worldwide and maintain a public register of every DP certificate issued.

The certification of Engineering Officers is provided on satisfactory completion of an HV course.

Experience

Key DP Personnel	Any DP Vessel		Present DP Vessel	
	Hours	Weeks	Hours	Weeks
Master / OIM	250	10	100	4
Senior DPO	250	10	150	2
Junior DPO	150	3	50	1
Chief Engineer	250	10	100	4
ECR Engineer	100	4	50	2
ETO / ERO	250	10	100	4
Electrician	250	10	100	4

Figure 10.14

IMCA Recommended Experience Levels – Existing Vessels

With a new vessel or management, the following minimum levels should be maintained for the initial period of six months, at which point the standard levels of experience should be adhered to.

Key DP Personnel	Minimum vessel experience	
Master / OIM	50 hrs	Over 7 days at sea
Senior DPO	50 hrs	Over 7 days at sea
Junior DPO	50 hrs	Over 7 days at sea
Chief Engineer	21 days	Including 7 at sea
ECR Engineer	14 days	Including 7 at sea
ETO / ERO	21* days	Including 7 at sea
Electrician	21* days	Including 7 at sea

Figure 10.15

IMCA Recommended Experience Levels – New Vessels

* If an ETO /ERO and an electrical engineer are normally onboard at the same time, the vessel time can be reduced to 14 days each.

International Marine Contractors Association (IMCA)

The International Marine Contractors Association (IMCA) represents offshore, marine and underwater engineering companies world-wide. Formed in 1995 from the amalgamation of the Dynamic Positioning Vessel Owners Association (DPVOA) and the International Association of Offshore Diving Contractors (AODC),

IMCA M117

Training and Experience of Key DP Personnel

IMCA M117 provides guidance on the levels of training and experience expected for any Dynamic Positioning Officers or Engineering Officers onboard DP Vessels.

Reference

IMCA is dedicated to the promotion of its members' common interests and the provision of a single authoritative voice for its members.

As the range of activities of member companies is so extensive, IMCA is divided into four separate divisions, each covering a specific area of operation. The divisions are Diving, Marine, Offshore Survey and Remote Systems & ROV.

- The Diving Division covers safe practice in the offshore diving industry, including all aspects of equipment, operations and personnel related to offshore diving operations (including atmospheric diving systems).
- The Marine Division covers all aspects of vessel operations and marine equipment, with particular focus on dynamic positioning related issues.
- The Offshore Survey Division is concerned with all aspects of equipment, operations and personnel relating to offshore survey operations.
- The Remote Systems & ROV Division is concerned with all aspects of equipment, operations and personnel related to remotely controlled systems (including ROVs) used in support of marine activities.

In addition, all members benefit from the work of the two international core committees, receiving regular briefing and being represented with client and regulatory bodies at an international level across the broad range of members' operations. These two core committees are concerned with:

- Safety, Environment and Legislation
- Training, Certification and Personnel Competence

These four divisions and two core committees are overseen by the Associations 'Overall Management Committee' which comprises the chairmen and vice-chairmen of the divisional and core committees along with the Chief Executive and Technical Director.

As such, the activities of IMCA are broad in scope whilst being based on sound principles of knowledge, experience and openness in a variety of disciplines.

Although separate entities within the association, a high degree of cross communication ensures that lessons learnt by one group is disseminated to the other groups.

The purpose of IMCA can be categorised as follows:

- The provision of standards with regard to equipment, operations and personnel for Dynamic Positioning Operations.
- The cross-communication of information, advice and guidance from one contractor or operator to another.

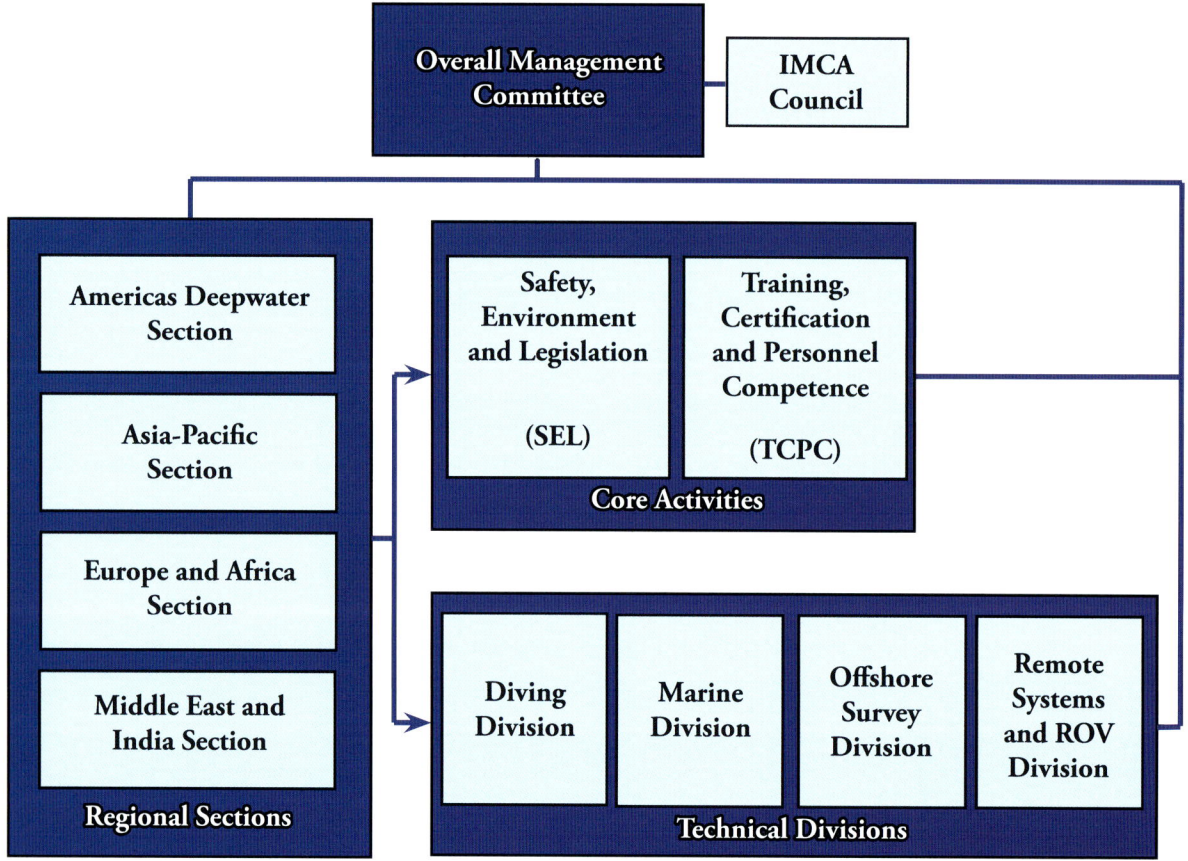

Figure 10.16   IMCA Organisation

- The advancement of techniques and operational functionality of Dynamic Positioning Systems.

In order to provide the required level of guidance and advice, IMCA provide a number of means for dissemination and publication of information.

Information Notes

These are issued as and when required and contain information that can be explained or described within a very short narrative. Often these notes will be the precursor to a more detailed report.

Safety Flashes

Important safety related information is provided as safety flashes. These are provided to all members of IMCA and are issued as soon as possible after an incident or the reporting of applicable advice.

Reports

Detailed advice and guidance are issued as reports, usually written by third party contractors specialised in the particular field being investigated. These reports contain information, advice and recommendations for equipment, personnel and operational guidelines.

Standard Reporting Formats

In order to provide cross-company and cross-industry dissemination of information, standard reporting formats are provided by IMCA for Station Keeping Incidents (DP) and Lifting Appliance Incidents. IMCA provide summaries of yearly incident reports and are working on a number of other items (such as incident databases).

Seminars

IMCA hold an annual seminar for each division of the association. These are sometimes also linked to another division to assist with the inter-divisional flow of information.

Membership of IMCA is to be encouraged for all companies operating Dynamically Positioned Vessels in order to ensure that the Officers that are required to utilise the equipment, are provided with all relevant safety and operational advice that has been collated from the vast array of IMCA member companies.

Chapter 11

THE ISM CODE

General Introduction

The International Safety Management (ISM) Code was adopted by the International Maritime Organization in May 1994 as a new Chapter IX to the SOLAS Convention. The requirements of the ISM Code entered into force as indicated below:

All Passenger Ships, including Passenger high speed light craft (regardless of size)	1st July 1998
Oil Tankers, Chemical Tankers, Gas Carriers, Bulk Carriers and Cargo High Speed Light Craft of 500 gross tonnes and above	1st July 1998
Other Cargo Ships and Mobile Offshore Drilling Units of 500 gross tonnes and above	1st July 2002

The establishment of the ISM Code can be tracked back to the incidents surrounding the capsizing of the *Herald of Free Enterprise* where failures in the onboard and shore based management of the vessel and the ship to shore link, resulted in the loss of many lives. The requirements of the ISM Code are to ensure that critical safety concerns are addressed and that sufficient support is provided by the shore based management to the vessel management team.

The ISM Code therefore establishes an international standard for the safe operation of ships by providing a structured Safety Management System (SMS) based on recognised safety management principles.

The objectives of the code are to ensure safety at sea, prevention of human injury or loss of life, and avoidance of damage to the marine environment, and to property.

The Safety Management System should:
- Provide for safe practices in ship operation and a safe working environment.
- Establish safeguards against all identified risks.
- Continuously improve safety management skills of personnel ashore and onboard, including preparing for emergencies relating both to safety and environmental protection.
- Ensure compliance with mandatory rules and regulations.

Definitions

The following definitions are provided for the common terminology associated with the ISM Code.

Audit (Internal)

An audit arranged and carried out by the company to evaluate the implementation and effective functioning of its safety management system both ashore and afloat. The company should carry out an internal audit of the office and each of its vessels at least once a year.

Auditee

An organisation, department or ship to be audited.

Auditor

A qualified person who is authorised to carry out an audit.

Designated Person Ashore (DPA)

A person or persons having direct access to the highest level of management and whose responsibility and authority includes monitoring the safety and pollution prevention aspects of the operation of each ship and ensuring that adequate resources and shore based support are applied as required.

Document of Compliance (DOC)

Certificate issued to a company which complies with the requirements of the ISM Code.

Finding (FN)

A non-fulfilment of objectives or requirements defined by the company, which goes beyond what should be subjected to mandatory ISM Code Certification. It does not affect the issue or renewal of a DOC or SMC.

Incident (Hazardous Occurrence)

An undesired event which, under slightly different circumstances, could have resulted in an accident.

Lead Auditor

An auditor who is qualified and authorised to manage an audit.

Major Non-Conformity (MNC)

An identifiable deviation which poses a serious threat to personnel or ship safety or serious risk to the environment and requires immediate corrective action. The lack of effective and systematic implementation of a requirement is also considered as a major non-conformity.

Non-Conformity (NC)

An observed situation where objective evidence indicates the non-fulfilment of a specified requirement.

Objective Evidence

Quantitative or qualitative information, records or statements of fact pertaining to safety or to the existence and implementation of a SMS element, which is based on observation, measurement test, and which can be verified.

Observation (OBS)

A statement of fact made during a safety management audit and substantiated by objective evidence. It may also be a statement made by the auditor referring to the SMS which, if not corrected, may lead to a non-conformity in the future (may also be applied for a proposal for improvement).

Safety and Environmental Policy

A clearly-worded statement summarising the company's safety management objectives and describing how they will be achieved. The policy should be signed by the Chief Executive (or equivalent) and reviewed at regular intervals.

Safety Management Audit

A systematic and independent examination to determine whether the SMS activities and related results comply with planned arrangements and whether these arrangements are implemented effectively and are suitable to achieve objectives. It is required to evaluate the effectiveness of the SMS and to identify areas that need improvement.

Safety Management Certificate (SMC)

A document issued to a ship which signifies that the company and its shipboard management are operated in accordance with the approved safety management system.

Safety Management System (SMS)

A ' Safety Management System' means a structured and documented system enabling personnel to effectively implement the company's safety and environmental protection policy. It must include all of the specific requirements of the ISM Code, but the procedures themselves should be tailored to reflect the activities of the company (e.g. vessel type, trade, operations) and any potential hazards that may be anticipated. The system may incorporate other relevant material by reference (e.g. Industry publications, circulars) but if so, such documentation should always be kept up to date.

System Review (Management / Masters Review)

A formal evaluation by the top management of the status and adequacy of SMS and need for improvements.

Functional Requirements of the ISM Code

In order to fulfil the requirements of the ISM Code, every company responsible for the management of a vessel must develop, implement and maintain a Safety Management System which includes the following functional requirements:

- A Safety and Environmental Protection Policy.
- Defined levels of authority and lines of communication for shore and shipboard personnel, and the provision of a link between the two.
- Instructions and procedures to ensure safe operation of ships and protection of the environment in compliance with relevant international flag State legislation.
- Procedures for reporting accidents and non-conformities.
- Procedures to prepare for and respond to emergency situations.
- Procedures for internal audits and management reviews.

A Safety Management System meeting the requirements of the ISM Code therefore requires a company to:

- Document its safety management procedures.
- Record its actions to ensure that conditions, activities and tasks that affect safety and the environment are properly planned, organised, executed and checked.

ISM Code Certification

ISM Code certification can be divided into the requirements for onshore management and for the shipboard management.

Onshore Management

A 'Document of Compliance' (DOC) will be issued to the onshore management group, on completion of an external audit to confirm compliance with the requirements of the ISM Code. This external audit will be conducted by an organisation recognised by the vessel's flag State administration. The certificate will define the specific vessel types that the company are deemed acceptable to manage.

The 'Document of Compliance' is valid for a period of five years with annual external verification by the recognised authority.

A copy of the certificate should be held onboard all company vessels that the managers are responsible for under the requirements of the ISM Code.

It is important to note that the withdrawal of the 'Document of Compliance' has widespread consequences. If the 'Document of Compliance' of a company that manages ten vessels is retracted or suspended, the entire fleet of vessels will also be affected with the retraction or suspension of each individual 'Safety Management Certificate'.

If, alternatively, the 'Safety Management Certificate' of one of the ten vessels were retracted or suspended, the 'Document of Compliance' would remain unaffected and the managers and remaining vessel fleet would be able to continue to operate.

Shipboard Management

A 'Safety Management Certificate' (SMC) will be issued to each individual vessel, on completion of an external audit to confirm compliance with the requirements of the ISM Code. This external audit will be conducted by an organisation recognised by the vessel's flag State Administration. The audit will be required to ensure that the company and the shipboard management operate in accordance with their authorised SMS.

The 'Safety Management Certificate' is valid for a period of five years and will be subject to a minimum of one intermediate audit between the second and third anniversaries of the initial audit.

Safety and Environmental Policy

In order to describe how the objectives of the SMS are to be achieved, a 'Safety and Environmental Protection Policy' must be established by the vessel managers.

The forms that such a policy may take will vary from company to company, however the basic content will be very similar in nature, with the three following fundamental concerns:

- Safety of Personnel
- Safety of the Vessel
- Safety of the Environment

Defined Levels of Authority and Lines of Communication

To ensure the safe operation of each vessel and to provide a link between the onshore management and those onboard each vessel, the vessel managers must have a 'Designated Person Ashore' (DPA).

The responsibility and authority of the DPA must include monitoring the safety and pollution prevention aspects of the operation of each vessel and to ensure that adequate resources and shore based support are applied, as required. To ensure that these requirements are achieved, the DPA must have direct access to the senior level of management.

The link between the offshore management team, primarily the Master must be clearly defined and similarly, the Master's responsibilities and overriding authority must be unambiguously stated.

Master's Responsibilities

The Vessel Master is responsible for:

- Implementing the relevant company policies relating to safety and environmental protection.
- Motivating the crew in the observation of that policy.
- Implementing the relevant company policies relating to safety and environmental protection.
- Issuing appropriate orders and instructions in a clear and simple manner.
- Verifying that specified requirements are observed.
- Reviewing the SMS and reporting its deficiencies to the shore based management.

Master's Authority

The Master has the overriding authority and responsibility to make decisions with respect to safety and pollution prevention and to request the company's assistance as necessary.

An example of a 'Safety and Environmental Policy' may include statements such as:
- The company is committed to the elimination of harmful incidents onboard all vessels.
- The company consider all accidents, dangerous occurrences, occupational illnesses and environmental incidents to be preventable.
- The company will promote a positive Health, Safety and Environmental culture throughout all onshore and offshore sites.
- The company shall develop a high degree of Health, Safety and Environmental Awareness and Competence in all employees.
- The company expect the same level of Health, Safety and Environmental standards from all third party contractors with whom we interact and contract.

Figure 11.1 Example Safety and Environmental Policy

Onboard Management

In addition to the responsibilities and authorities of the onshore management team and the Master, defined job descriptions, including responsibilities and reporting lines must be provided for all safety critical positions.

Instructions and Procedures

In order to ensure that the key operations onboard the particular type of vessel are conducted in the correct manner, procedures must be in place for the identified operations that the vessel will routinely conduct. For Offshore Support Vessels these operations would include as a minimum the following:

- Familiarisation
- Navigation and Bridge Procedures
- Pilotage and Pilot Boarding
- Engine Room Procedures
- Crane, Cargo, Diving, ROV, Survey or Pipe Lay Operations
- Mooring
- Gangway Use, Security and Piracy
- Ballasting, Stability and Watertight Integrity
- Maintenance
- Surveys and Tank Inspections
- Bunkering
- Potable Water
- Personnel Basket Transfer
- Helicopter Operations
- Adverse Weather Policy
- Waste Management
- Training
- Incident Reporting

Reporting Accidents and Non-Conformities

A method must be in place for the reporting of accidents, incidents and non-conformities. Once reported by the vessel the details must be investigated and analysed with the objective of improving safety and pollution prevention.

Non-conformance Reporting and Corrective Action

Non-conformances may be reported by the onshore management team, an internal or external auditor or following a Master's or Management Review.

The corrective actions resulting from the non-conformance should be implemented to help prevent a possible incident or accident resulting from the original non-conformance.

Accident and Incident Reporting

The initial reporting and investigation of an incident is the responsibility of the onboard management team, however the reporting of the incident to the onshore management team must be completed at the soonest opportunity to ensure that full support can be provided.

The most important part of any accident and incident reporting will be the preventative actions that should be completed to ensure that the possibility of a similar occurrence is limited as far as is possible.

Emergency Situations Procedures

The vessel managers must develop and implement processes and procedures for responding to various emergency situations. In order to monitor the effectiveness of the emergency procedures, emergency drills should be performed on a periodic basis with any lessons learned from these drills transferred to the procedures. In certain companies, a series of emergency checklists have been developed for onboard assistance in planning of shipboard exercises and to assist in real emergency situations.

Audits and Reviews

The SMS is subject to periodic audits to verify whether safety and pollution prevention activities comply with the requirements of the ISM Code. In addition, internal audits are used to determine the effectiveness of the SMS, to optimise costs, and to identify areas for improvement.

The audits and reviews of the SMS can be identified as the following:

- External Audit (Classification Society)
- Internal Audits
- Master's Reviews
- Management Reviews

External Audits (Classification Society)

For office sites, a periodical verification of the 'Document of Compliance' will be carried out to maintain the validity of the certification. This external audit will be conducted within three months before and after each anniversary date of the 'Document of Compliance'.

For vessels, an intermediate verification of the 'Safety Management Certificate' will be carried out to maintain the validity of the certification. This external audit will be conducted between the second and third anniversary date of the issue of the 'Safety Management Certificate'.

Internal Audits

The completion of the ISM Code Internal Audit will constitute a series of interviews with vessel staff and an examination of the vessel's documentation and records. The objective of the Internal Audit is to examine the SMS onboard and to determine (by objective evidence) the degree of compliance with the applicable rules, regulations and standards, and the documented procedures, also reflecting company requirements.

The audit should identify strengths and weaknesses, verify that procedures are controlled properly and that they are being followed. The system audit should also establish corrective action and improvement plans and ensure that personnel understand the requirements and that management commitment is evident. Results of the Internal Audit, positive and negative, will be a part of the management review in the continuous improvement

In order to ensure that the requirements of the ISM Code are fully reviewed, the Internal Audit should be divided into manageable sections covering all aspects of the vessels operation, such as:

- Safety and Environmental Protection Policy
- Company Responsibilities and Authority
- Designated Person
- Master's Responsibility and Authority
- Resources and Personnel
- Development of Plans for Shipboard Operations
- Emergency Preparedness
- Reports and Analysis of Non-Conformities, Accidents and Hazardous Occurrences
- Maintenance of the Ship and Equipment
- Documentation and Publications
- Company Verification, Review and Evaluation

Master's Reviews

Formal Master's reviews of the SMS are carried out at periodic intervals to ensure its continuing suitability, adequacy and effectiveness in satisfying the requirements of the ISM Code and the stated policies and objectives. Master's reviews should consist of an evaluation of the following:

- Most recent
- Findings of audits
- Documented analysis of accidents, incidents and hazardous situations
- The overall effectiveness of the SMS in achieving stated objectives
- Feedback information
- Maintenance system

Management Reviews

Formal Management Reviews of the SMS should be periodically conducted to ensure the continuing suitability, adequacy and effectiveness of the SMS in satisfying the requirements of the ISM Code andthe stated policies and objectives. Management Reviews should consist of an evaluation of the following:

- Follow-up to the last Management Review
- Most recent Master's Review from each vessel
- Policy and objectives
- Findings of audits
- Feedback information
- Documented analysis of accidents, incidents and hazardous situations
- Recommendations and memos following class and statutory surveys
- The overall effectiveness of the SMS in achieving stated objectives
- Considerations for up-dating the system in relation to changes in fleet, trade, new regulations, market strategies and environmental conditions and requirements
- Maintenance system.

Figure 11.2 Olympic Canyon – ROV Support Vessel

Chapter 12

THE ISPS CODE

General Introduction

The International Ship and Port Facility Security (ISPS) Code, implemented under SOLAS 1974, came into effect on 1st July 2004. Instigated following the events in New York on 11th September 2001, the Code sets out to provide effective security measures for ships, ports and the states that they operate within. The Code has had significant implications for all vessel types and shipping companies and for this reason, a brief summary of the principles and requirements of the Code are provided below.

The Code has been implemented as Chapters XI-1 and XI-2 of SOLAS. Part A of the Code details the mandatory requirements required to be in place in order to comply with the Code, with Part B providing guidance on how shipping companies and vessels can ensure compliance with the requirements of the Code.

Although the requirements of the ISPS Code apply to all vessel types, it is significant that the regulations have been extended to all offshore installations. As the regulations not only affect Offshore Support Vessels directly but also indirectly as all ports and offshore drilling rigs and platforms are also required to comply, details provided here may be of particular relevance.

Failure to comply with the Code may lead to vessels being detained in port or excluded from entering port. With the principle of the Code being to maintain security, non compliance could however leave the vessel more vulnerable to terrorist attacks. Whether compliance would reduce the possibility of such an attack is obviously open to debate.

ISPS Code Abbreviations and Definitions

The following abbreviations and definitions are provided for the common terminology associated with the ISPS Code.

Company Security Officer CSO

Every shipping company and offshore installation operator is required to appoint a CSO. The CSO is the onshore person responsible for the development, implementation and maintenance of the Ship Security Assessment and Ship Security Plan.

International Ship and Port Facility Security Code ISPS

The ISPS Code was implemented on 1st July 2004. The Code details the requirements for onshore and offshore security measures to be maintained within the shipping industry following the terrorist attacks of 11th September 2001.

Port Facility Security Assessment PFSA

All Port Facilities are required to comply with the ISPS Code and are required to conduct a Port Facility Security Assessment to detail any further requirements and actions necessary to comply with the Code.

Port Facility Security Officer PFSO

Every Port Facility is required to appoint a Port Facility Security Officer. The PFSO is responsible for the development, implementation and maintenance of the Port Facility Security Assessment and Port Facility Security Plan.

Port Facility Security Plan PFSP

Every Port Facility is required to develop a Port Facility Security Plan outlining levels of security alert and security measures in place.

Recognised Security Organisation RSO

Government authorities can authorise recognised security organisations to conduct the required verifications and approvals on their behalf. The relevant government authorities must therefore approve the security organisations.

Ship Security Assessment SSA

Every vessel is required to comply with the ISPS Code and are required to conduct Ship Security Assessments to detail any further requirements and actions necessary to comply with the Code.

Ship Security Officer SSO

Every vessel is required to have a SSO. The SSO is the offshore responsible person for implementation and maintenance of the Ship Security Plan for that particular vessel.

Ship Security Plan SSP

Every vessel is required to develop a Ship Security Plan outlining security measures in place.

Functional Requirements of the ISPS Code

The practical requirements of the Code include the need for all vessels to display a vessel specific identity number on the ship's side, install a security alert system, be fitted with a ship identification system and to keep a daily synoptic report detailing the vessels activities.

Further requirements of the Code include the need to control the access of persons to the vessel, to restrict access to certain parts of the vessel and to screen personal baggage.

All vessels are required to conduct a vessel specific Ship Security Assessment (SSA). This assessment will be used as the basis for the provision of a Ship Security Plan (SSP) approved by the flag State Authority or by a recognised Security Organisation (RSO). The approved SSP must be implemented onboard the vessel, verified and records of drills maintained.

In order to implement and continually develop the Ship Security Assessment and Ship Security Plan, each company will be required to appoint a Company Security Officer (CSO) and a Ship Security Officer (SSO) for each vessel within their particular fleet.

ISPS Code Certification

Every vessel over 500 gross registered tonnes to which the requirements of the ISPS Code apply shall require to be issued with an International Ship Security Certificate which will be valid for a period of 5 years.

In order to comply and maintain such certification every vessel to which the requirements of the ISPS Code apply shall be subject to:

- An initial verification prior to the issue of an International Ship Security Certificate. This verification shall ensure that the vessels Ship Security Plan, security equipment and awareness and training onboard all comply with the requirements of the ISPS Code.
- A renewal verification at intervals not exceeding 5 years. This verification shall ensure the continued applicability of the Ship Security Plan, maintenance of security equipment, training records and security awareness.
- An intermediate verification, to take place between the second and third anniversary date of the initial issue of the International Ship Security Certificate. This verification shall ensure the continued applicability of the Ship Security Plan, maintenance of security equipment, training records and security awareness.

Company Security Officer

Every company is required to appoint a Company Security Officer (CSO). This CSO may be appointed for a single vessel or a number of vessels and it is more than acceptable for each company to appoint more than one CSO, provided that the vessels for which the CSO has responsibility, is clearly defined.

In accordance with the requirements of the Code, the duties and responsibilities of the CSO should include, as a minimum:

- Advising the level of threats likely to be encountered by the vessel, dependent on the vessels location, local terrorist activities and governmental advice.
- Ensuring that appropriate and thorough Security Assessments are conducted.
- Ensuring that the Security Plan is developed, approved by the appropriate authority and that it is implemented onboard and is subject to continuous improvement.
- Ensuring that regular reviews and audits of the Ship Security Plan and all onboard ship security activities are conducted.
- Ensuring that any deficiencies noted during training, drills or internal reviews are rectified in a timely manner.
- Ensuring that security awareness is promoted onboard and that all onboard are aware of the vigilance required at all times.
- Ensuring that all onshore and offshore personnel are provided with the relevant training required in accordance with the Code.

Ship Security Officer

The Code requires that a Ship Security Officer (SSO) is appointed for each vessel.

In accordance with the requirements of the Code, the duties and responsibilities of the SSO should include, as a minimum:

- Completing regular security inspections of the vessel to ensure that the appropriate security measures, detailed in the vessels Ship Security Plan, are maintained.
- Implementation of the Ship Security Plan onboard the vessel.
- Improvement of the Ship Security Plan as necessary.
- Ensure that close communication with the Company Security Officer is maintained and that any deficiencies in the Ship Security Plan are reported to the Company Security Officer.
- Ensure that appropriate training is provided to all onboard personnel and that onboard vigilance is maintained.
- Ensure that the appropriate security equipment is onboard and available for use.

Ship Security Assessment

The IMO require the Company Security Officer to

ensure that a Ship Security Assessment is conducted prior to the development of a Ship Security Plan. This assessment is required to ascertain any areas where security measures may require improvement and to ensure that the development of the Ship Security Plan encompasses all relevant areas of the vessel and its operation.

As a minimum, the Ship Security Assessment should include an assessment and verification of the vessels:

- Physical security, including all areas of dedicated or potential access. This is of particular importance on Offshore Support Vessels with low freeboards and after decks and often open sterns.
- Operational characteristics. For this type of vessel this could include the potential for increased risks due to standard 'runs' for vessels between certain ports and offshore sites. In addition the vessels transit speed tends to be, relative to other shipping, low and the vessel type tends to be in one static location for prolonged periods during offshore operations.
- Structural integrity of the vessel.
- Personnel protection systems.
- Shipboard procedures in place for onboard security whilst at sea and in port.
- Communications systems in place on board the vessel, both internal and external.

Figure 12.1 Subsea Viking – Offshore Support Vessel

Ship Security Plan

The IMO require the Company Security Officer to ensure that a Ship Security Plan is developed for each company vessel. This plan is required to detail the security measures in place onboard the vessel.

As a minimum, the Ship Security Plan should include the following:

- Details of the onboard and onshore security organisation provided for the company and the vessel.
- Details of the communications systems in place to maintain onboard communications and exterior communications with port facilities, other vessels and the company.
- Details of the security measures in place for all levels of security.
- Details of the required reviews, audits and processes in place to ensure that the plan is continually improved.
- Details of reporting procedures to the appropriate authorities and company contacts in the event of an incident relating to vessel security.

On completion of the Ship Security Assessment and the subsequent development of the Ship Security Plan, the Ship Security Plan should be submitted to the relevant flag State Authority or an approved Recognised Security Organisation (RSO) for approval on behalf of the flag State Authority.

Training, Drills and Exercises on Ship Security

With regards to the ISPS Code, training can be divided into three distinct parts, namely training for the Company Security Officer (and appropriate shore based personnel), Ship Security Officer and for the onboard personnel.

The Company Security Officer (and any appropriate shore based personnel), should be provided with appropriate training to ensure full awareness and familiarity with the Code, knowledge of applicable flag State requirements, knowledge of the means of performing Ship Security Assessments and the development of Ship Security Plans and skills required for the continuous maintenance and improvement of the Ship Security Plan.

The Ship Security Officer should, in addition to the training required for the CSO above, be provided with appropriate training in crowd management, vessel familiarisation, use and maintenance of security equipment and familiarisation in the Ship Security Plan and associated Emergency Response procedures.

Shipboard personnel who may be delegated specific responsibilities and functions in the Ship Security Plan should also receive training in their designated tasks, including knowledge of security levels and current threats, crowd management techniques, knowledge of Emergency Response procedures and awareness of security measures.

Full details of the training required are contained in SOLAS and the ISPS Code and in The Nautical Institute book *Maritime Security*.

Subsea Viking – Offshore Support Vessel

Chapter 13

SHIPBOARD SAFETY

General Introduction

Shipboard safety is a wide ranging and expansive subject that could feasibly include aspects of all operations conducted onboard a ship. Many shipboard tasks have elements of associated risk and therefore an element of risk assessment is fundamental to the completion of these tasks in a safe and controlled manner. Although the type of task may vary, all shipboard operations can be completed in a safe manner by ensuring that a controlled system of work is in place, in a safe environment and completed by trained and experienced personnel. Such systems can therefore be applied to any task and it is this overall general safety structure and system that will be concentrated on within this section. In addition, specific high risk or high frequency activities and the associated safety considerations will be examined.

This section has therefore been divided into the following safety related subject headings:

- Merchant Shipping (Health and Safety at Work) Regulations
- The Principles of Health and Safety
- Human Factors and the Human Element
- Safety Leadership
- Shipboard Safety Organisation
- Safety Officers
- Safety Representatives
- Safety Committees and Safety Meetings
- Safety Area Inspections
- Accident and Incident Reporting
- Accident and Incident Investigations
- Inductions
- Safe Access and Slips, Trips and Falls
- Risk Assessments and Hazard Identification
- Permit to Work Systems
- Entry into Enclosed Spaces
- Mooring Operations
- Manual Handling
- Lifting Operations and Lifting Equipment
- Provision and Use of Work Equipment
- Working at Height

Merchant Shipping (Health and Safety at Work) Regulations

The Merchant Shipping and Fishing Vessels (Health and Safety at Work) Regulations 1997 have been in force since 31st March 1998. These regulations apply to vessels registered in the United Kingdom and to all vessels when operating in United Kingdom waters. Therefore, irrespective of the port of registry of the vessel, the regulations apply to any Offshore Support Vessel when operating in United Kingdom waters.

As the main function of the regulations is to ensure that employers protect the health and safety of all personnel engaged in operations onboard their vessels as far as is reasonably practicable, for the purposes of this section, it is assumed that these regulations apply and that best practice is followed.

Who is Responsible?

The ownership, management and crewing of any ship can often include three different companies with third party sub-contractors also a consideration. The Master, for example, may not work directly for the vessel owner or management company but may be employed via a crewing agency and his officers and crew may work for different crewing agencies entirely. It can therefore be difficult to define responsibilities with regards to safety onboard the vessel as numerous 'employers' may be involved. However, any employer and any management company have responsibilities that can be summarised as below.

Duties of the 'Employer'

The Health and Safety at Work Regulations place specific duties on the employer of the Master, officers and crew to provide:

- A safe working environment.
- Safe equipment, systems and machinery.
- Suitable and sufficient training, instruction and supervision on all equipment, systems and machinery onboard.
- Relevant guidance and information on the use and maintenance of specific equipment, systems and machinery.
- Appropriate personal protective equipment for the operations that may feasibly be conducted onboard the vessel.
- A system for onboard risk assessment.
- Health surveillance of workers, if appropriate.
- Consultation and discussion between the shipboard management and onboard elected safety representatives on all health and safety issues.

Duties of the Management Company

The management company has similar responsibilities to that of the employer, irrespective of whether the management company is the direct employer. The management company must assess the risks to any workers onboard and is therefore required to:

- Consult and co-ordinate with all other employers onboard the vessel about the health and safety of their employees and the measures to be taken to ensure their continued health.
- Provide information to employees about ship systems, equipment and machinery.
- Appoint a shipboard safety officer.
- Organise the election of safety representatives and the formation of a shipboard safety committee.

Duties of Employees

Irrespective of the duties which the Health and Safety at Work Regulations place on employers and management companies, safety must never be considered a management level responsibility alone. The regulations dictate duties on all employees to:

- Ensure that they pay reasonable care and attention to their own health and safety and that of all other personnel onboard the vessel that may be directly or indirectly affected by their actions or failures.
- Actively co-operate with all personnel onboard who are involved in health and safety duties and activities including the enforcement of any control measures which have been identified during risk assessments.
- Report any identified hazards or deficiencies immediately to the safety officer or other appropriate person.
- Utilise equipment and systems correctly in accordance with the relevant procedures and manufacturer's instructions.
- Treat any hazard to health or safety with due caution.

The duty of the employee is an aspect of all shipboard operations which cannot be stressed enough. Every single crew member onboard has a duty to rectify or report any deficiency or situation which they feel may affect their own or the safety of others. The accompanying text (see box below) is a well known poem which has been circulated throughout the industry for many years and acts as a valid reminder as to how important the employee can be in ensuring the safety of his fellow crew.

The Principles of Health and Safety

In order to provide a suitable framework for ensuring the safety of the crew and any third party personnel onboard, the Health and Safety at Work Regulations are based on a series of seven principles. These seven principles are integral to the safe operation of any site or vessel and must all be present in order to fulfil the provision of a safe working environment, combining together to form an integrated and coherent system of work.

I CHOSE TO LOOK THE OTHER WAY

I could have saved a life that day, but I chose to look the other way
It wasn't that I didn't care, I had the time, and I was there
But I didn't want to seem a fool, or argue over a safety rule

I knew he'd done the job before, if I called it wrong, he might be sore
The chances didn't seem that bad, I've done the same and he knew I had
So I shook my head and walked on by, he knew the risks as well as I
He took the chance, I closed an eye, and with that act, I let him die

I could have saved a life that day, but I chose to look the other way
Now every time I see his wife, I'll know I should have saved his life
That guilt is something I must bear, but it isn't something you need share

If you see a risk that others take, that puts their health or life at stake
The question asked, or thing you say, could help them live another day
If you see a risk and walk away, then hope you never have to say
I could have saved a life that day, but I chose to look the other way

Author: Unknown

	Principles
1	Avoidance of Risk
2	Evaluation of unavoidable risk
3	Procedures and working environment
4	Adaptation of procedures
5	Safety management system
6	Protection measures
7	Appropriate and relevant information

Figure 13.1 Principles of Health and Safety

Avoidance of Risk

For any operation to be performed onboard the vessel, the risks associated with the operation should, if possible, be completely removed. This can be accomplished by removing the risk at the source or by adopting an alternate means of completing the task by using different procedures, equipment, systems or substances.

Evaluation of Unavoidable Risk

All operations performed onboard the vessel should be assessed prior to commencement of the task. Any unavoidable risks associated with the particular task should be evaluated and any actions required to reduce these risks should be identified.

Procedures and Working Environment

Procedures in place and the working environment for any tasks should be developed to alleviate monotonous and repetitive work. The working environment should consider the location of the work, external factors affecting the work (such as the weather conditions) and importantly the crew members' capabilities. The procedures, equipment, systems and personnel should suit the task in hand.

Adaptation of Procedures

Procedures must be continually reviewed and updated. These reviews should consider any changes to equipment, systems, the working environment or onboard working practices that could adversely affect the health and safety of the crew involved in the task and those who may be indirectly affected.

Safety Management System

A structured and consistent safety management system and onboard safety organisation is crucial to the health, safety and welfare of those onboard. An SMS should act as a framework for safe work onboard the vessel to ensure that the actions of any individual cannot adversely affect the health and safety of that individual or any other person onboard the vessel.

Protection Measures

Protection measures can take the form of personal protective equipment (PPE) such as coveralls, hard hats, gloves, safety boots, ear defenders and eye protection. However, other protective measures can include equipment guards, safety barriers and electrical isolations. Such protective measures are integral to safe working and the equipment necessary to provide such protection must be provided onboard.

Appropriate and Relevant Information

Any guidance, information or instructions relevant to a particular task or equipment or systems to be utilised to fulfil the task must be made available to those planning and completing the task. Any limitations or particular characteristics of the equipment or systems that may affect the task must be considered.

'Human Factors' and the 'Human Element'

The provision of a structured system which includes the principles of health and safety as described in the previous section is fundamental to the basis of a safe vessel. However there is a further element that must be in place in order for these systems to be followed in the correct manner; the human element.

In recent years a number of studies have been carried out regarding the human element in accidents. Within the marine industry, the UK P&I Club has been vigorously active in its determination not only to ascertain how many accidents are related to the human element, but also to consider why this is the case and how can we prevent this trend from continuing.

The chart shown below (figure 13.2) presents the results of detailed analysis the P&I Club published in 2003, which considered fifteen years worth of loss prevention incidents. A startling 62% were attributable to human error. Hence the common phrase, 'people cause accidents'.

Significant time and energy has been expended in the Club's efforts to analyse and quantify the reasons for such a high percentage of human element failures and their investigations have drawn on previous work by Shell (UK) and a number of leading United Kingdom Universities.

This research has resulted in the identification of eleven categories of latent failure that the P&I Club analysis suggests underlie all accidents.

Design

The poor ergonomic design of ship work stations, tools, systems and accommodation spaces can have a detrimental effect on the performance and hence the safety of the vessel crew. There are many instances where poor design can cause problems for the crew such as instances where a lack of sufficient rest periods is caused by adverse noise levels, the use of tools inadequately designed for the specific function onboard that particular vessel leads to a personnel injury or where the design of a manoeuvring console precludes 360 degrees visibility for the operator.

Hardware

Any equipment or ship's systems should be suitable for the intended use onboard the vessel. Quality, purpose built equipment, suitable for operation in the expected offshore environment is essential.

Maintenance

The correct design and suitability of any equipment

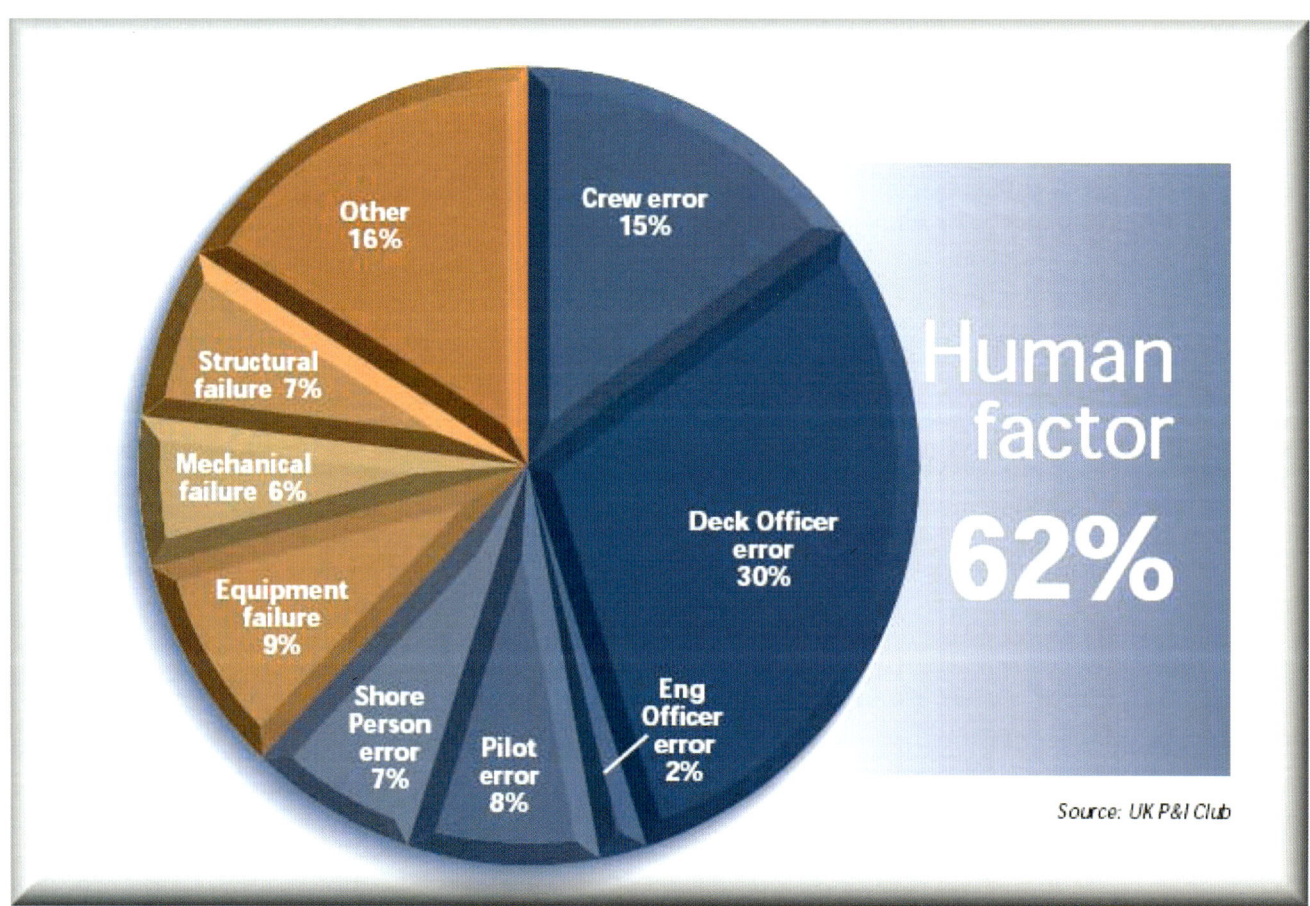

Figure 13.2 Human Factor Statistics

or ship's system will be of little relevance if the continued maintenance of such equipment and systems is not conducted. Poor maintenance regimes will obviously lead to more frequent and more serious machinery breakdowns.

Housekeeping

Poor housekeeping can lead to accidents, particularly slips, trips and falls and can breed a poor safety culture which may spread throughout the various departments onboard. Poor housekeeping is not only a safety issue, but also a health concern.

Error Enforcing Conditions

Physical conditions and other influences can have a negative effect on human judgement and reactions. This can include a variety of influences such as heat, cold, stress, noise or multiple distractions at the same time.

Procedures

Procedures alone do not make a safe system of work; however they do provide an important key function in any operation. The poor quality or lack of suitable procedures and instructions can lead to operations being conducted without due consideration of lessons learned from previous incidents and from recognised guidance and advice.

Training

Training in the use of onboard equipment, systems and health and safety related awareness can lead to failure to utilise equipment and systems correctly. Competence to complete the tasks allocated to employees is a significant failure. Competent people will assist in making the ship a safe environment.

Communication

Communication onboard a vessel is of particular importance to ensure that all departments are aware of the operations and activities currently ongoing and those planned to be conducted. This can be of particular importance during activities such as cargo operations at an offshore location where bulk discharges are being controlled from the engine room, but where direct communication with the platform or installation is via the wheelhouse.

Incompatible Goals

The modern shipping industry requires efficient operations. Offshore operations are expected to be completed as quickly as possible. There may therefore be times when the seafarer is expected to choose between optimum working methods based on the company or industry procedures and financial pressures.

Organisation

A clear and unambiguous structure of management, both onshore and onboard the vessel is essential to ensure that safety related issues are dealt with professionally and efficiently. The failure to have a coherent safety management organisation in place, may lead to failures in the identification and rectification of onboard hazards. However, a well structured safety management organisation will be ineffective unless the individuals are motivated and proactive towards safety related issues.

Defences

Defences must be in place to provide sufficient protection of people, the ship and the environment from the consequences from operational risks.

Alert

The Human Element has been given prominence in the quarterly bulletin *Alert* published by The Nautical Institute and sponsored by Lloyds Register. It is available free via the Institute's website www.nautinst.org.

IMO and the 'Human Element'

In addition to the research that has been undertaken by organisations such as the UK P&I Club, International Maritime Organization investigators have also concluded that the majority of shipping incidents have a clear and recognisable human element. Factors such as fatigue, poor communication, differences in culture between crew members, health, situational awareness, stress and isolation have all been identified.

In previous years the IMO has been regarded and in fact has regarded itself as a regulatory body providing instructions and regulatory guidance in order to enhance maritime safety. However, the increasing awareness that the human element is so critical in accident and incident prevention has resulted in the IMO reviewing their regime structure and to improve their efforts to support a more safety culture type of approach to shipping.

Resolution A.947 (23) Human Element Vision, Principles and Goals was adopted by the organisation in 2003 and proposes that the human element must incorporate the needs of the crew at the design stage and should be under continuous review throughout the lifetime of the vessel. As such, the training of not only seafarers, but also ship designers, surveyors and lecturers on the importance of the human element must also be considered fundamental.

The IMO Human Element Vision

The Vision
To significantly enhance maritime safety and the quality of the marine environment by considering human element factors in order to improve performance of ships crews through empowerment and a safety culture

The Principles
The human element is a complex multi-dimensional issue that affects maritime safety and marine environmental protection. It involves the entire spectrum of human activities performed by ships' crews, shore based management, regulatory bodies, recognized organizations, shipyards, legislators, and other relevant parties, all of whom need to cooperate to address human element issues effectively.

The Organisation, when developing regulations, should honour the seafarer by seeking and respecting the opinions of those that do the work at sea.

Effective remedial action following maritime casualties requires a sound understanding of human element involvement in accident causation. This is gained by a thorough investigation and systematic analysis of casualties for contributory factors and the causal chain of events.

In the process of developing regulations, it should be recognized that adequate safeguards must be in place to ensure that a "single person error" will not cause an accident through the application of these regulations.

Rules and regulations addressing the seafarers directly should be simple, clear and comprehensive.

Crew performance is a function of individual capabilities, management policies, cultural factors, experience, training, job skills, work environment and countless other factors.

Dissemination of information through effective communication is essential to sound management and operational decisions.

Consideration of human element matters should aim at decreasing the possibility of human error as far as possible.

The Goals
To have in place a structured approach for the proper consideration of human element issues for use in the development of regulations and guidelines by all committees and sub-committees.

To conduct a comprehensive review of selected existing IMO instruments from the human element perspective.

To promote and communicate, through human element principles, a maritime safety culture and heightened marine environment awareness.

To provide a framework to encourage the development of non-regulatory solutions and their assessment based upon human element principles.

To have in place a system to discover and to disseminate to maritime interests studies, research and other relevant information on the human element, including findings from marine and nonmarine incident investigations.

To provide material to educate seafarers so as to increase their knowledge and awareness of the impact of human element issues on safe ship operations, to help them do the right thing.

Safety Leadership

Safety leadership both onshore and offshore is fundamental to the effective and positive safety culture onboard a vessel. The leadership ashore can provide motivation to, primarily the vessel Master and delegated Safety Officer, and the Master can provide motivational impetus

lead by example and your officers and crew will follow

and man management on a day to day basis onboard. Without such strong focused leadership, the safety management system and shipboard organisation will be unable to sustain a good safety culture.

The Maritime and Coastguard Agency (MCA) has recognised this issue and have previously provided detailed guidance on core leadership skills that can be applied onboard to foster a positive outlook with regards to safety. Reference is therefore made to the MCA publication 'A Practical Guide for Leaders in the Maritime Industry', extracts of which are reproduced here. It should be emphasised that the following applies not only to the vessel Master and Offshore Manager, but can also be applied by the Deck and Engineering Officers, Shift Supervisors, Safety Officer or Safety Representatives.

The following Core Leadership Qualities have been identified:

Confidence and Authority

Instil respect and command authority

Lead the team by example

Draw on knowledge and experience

Remain calm in a crisis

Empathy and Understanding

Practise 'tough empathy'

Be sensitive to different cultures

Recognise the crew's limitations

Motivation and Commitment

Motivate and create a sense of community

Place the safety of the crew and passengers above everything

Openness and Clarity

Communicate and listen clearly

Respect and Command Authority

DO		DON'T
• have confidence in your decisions and stick to them • admit mistakes when you are sure you are wrong • demonstrate staff care and respect through everyday actions • earn respect through your actions • try to achieve better mutual ship-shore management understanding	**without authority and respect it is difficult for leaders to influence the behaviour of their crews, including safety related behaviour** 	• demand respect from subordinates • use the power vested in your position as a threat • refuse to listen when challenged • act unnecessarily tough when there is no justification • ignore shore based management • blame shore based management for the consequences of decisions

Lead the Team by Example

DO		DON'T
• be seen to follow simple, visible safety rules during everyday activities • be seen to be playing an active role • be seen to assist in subordinates tasks where necessary	**be seen to be practicing what you preach**  **pull your weight as a key part of the team**	• apply hard discipline for non-compliance whilst flouting the rules yourself • avoid getting your hands dirty with subordinates tasks

OFFSHORE SUPPORT VESSELS 123

	Draw on Knowledge and Experience	
DO		**DON'T**

DO
- ensure that you are up to date on safety requirements
- consider your strengths and weaknesses in people skills such as communication, motivation, team working, conflict resolution, crisis management, coaching and appraisal, discipline
- be prepared to acknowledge your own knowledge gaps and seek advice when you need to

Centre:
it is self-evident that adequate knowledge and expertise are pre-requisites for effective leadership

good knowledge of safety related regulations, codes and standards

experience and skills not only in technical and operational issues but also in people management

DON'T
- concentrate only on technical safety knowledge without considering people skills

---

	Remain Calm in a Crisis	
DO		**DON'T**

DO
- develop excellent knowledge of and confidence in the crew's abilities
- implement a firm policy on compulsory attendance at emergency safety training and response drills

Centre:
calmness in a crisis is particularly important in view of additional complications of different languages and nationalities that make up the crew which tend to be emphasised during emergencies

DON'T
- hold infrequent or inconsistent drills
- fail to address language issues in emergency planning

Practise Tough Empathy

DO		DON'T
• encourage the crew to provide feedback on their situation, feelings and motives, both in everyday situations and formally in pre-arranged communication sessions • be prepared to acknowledge, mirror or summarise feedback to demonstrate understanding, then to explain your conclusions and intended course of action	**tough empathy is important to convey to your crew that you understand their situation, feelings and motives, and to enable you as a leader to take the right courses of action whilst considering the views of the crew**	• listen to what people say but then take a different decision without clearly demonstrating that you have heard and understood their concerns • over emphasise 'listening' at the expense of 'decision making' — this can lead to a loss of respect and authority

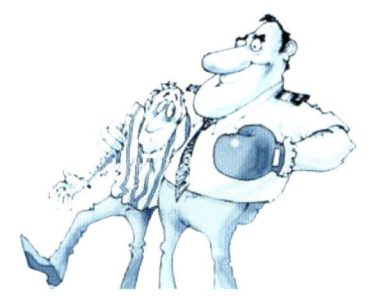

Be Sensitive to Different Cultures

DO		DON'T
• ensure that as far as is possible, one working language is used onboard and that the crew have adequate training in this language • try to avoid a critical mass of one nationality developing where possible • learn the key features of typical behavioural signals exhibited by the nationalities represented onboard • consciously seek to built trust, familiarity and integration of disparate social groups through organised social activities onboard	**it is important to know how to interpret different behavioural signals from different national cultures**  **it has been clearly demonstrated that different national cultures may have different values and attitudes towards safety**	• ingrain value judgements about different nationalities • overdo 'political correctness' in terms of dealing with different nationalities, so that relations become forced and unnatural

OFFSHORE SUPPORT VESSELS 125

	Recognise the Crew's Limitations	
DO		**DON'T**
• monitor and be aware of the signs of excessive fatigue in crew members • ensure that working hours are adequately supervised and recorded • discuss possible solutions for recurring problems with shore management • be able to decide when it is necessary to slow or halt operations temporarily	**good leaders have a clear understanding of how operational and other demands can realistically be met by the crew and are able to judge whether fatigue levels are such that action should be taken**	• rely on crew members to tell you if they are suffering from excess fatigue • accept that high levels of fatigue are an acceptable norm

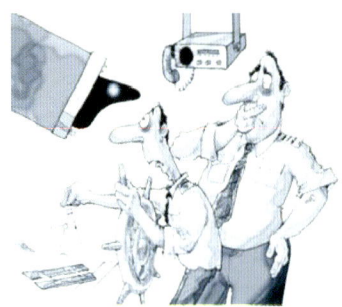

	Create Motivation and a Sense of Community	
DO		**DON'T**
• involve staff in aspects of management • ensure that feedback is always given on staff suggestions or questions • demonstrate interest in and care for crew welfare issues • take part in and encourage social activities involving the staff	**people in work are typically motivated by satisfaction or pride in completing a good job and a feeling of being part of a team. Morale has been shown to have an adverse effect/impact on error and violation rates; hence attention to these aspects is an important part of safety leadership**	• initiate one-off staff morale boosting initiatives or reward schemes that could be perceived as condescending or trivial • involve staff in theory, but in practice take little note of their input

	Place the Safety of the Crew and Passengers above everything	
DO		**DON'T**
	'nothing we do is worth getting hurt for ....' 	
• make it clear to both superiors and subordinates that you are empowered to act according to your own judgement on safety matters, without sanction from others • ensure that safety issues are integrated into other everyday operational activities, including walkabouts, meetings and one-to-one discussions		• declare that safety is your highest priority and then contradict this in your subsequent actions

	Communicate and Listen Clearly	
DO		**DON'T**
	clear two way communication and openness is necessary to achieve a 'just' culture where individuals feel free to speak up about problems or mistakes without being blamed. Balancing authority and approachability 	
• hold safety tours and informal discussions • ensure that your listening skills are adequate • implement an open door policy for crew members • ensure that there are no barriers preventing the open reporting of safety incidents and near misses • give positive feedback on what lessons have been learned through reporting of incidents and near misses • cultivate an atmosphere of openness through your own personal management style and everyday interactions		• hold safety tours which become primarily an excuse to check up on the crew and chastise them • declare a 'no-blame' policy without acknowledging the need for discipline • suggest schemes which are poorly followed up and maintained

Shipboard Safety Organisation

The implementation and continued maintenance of a positive safety culture onboard any vessel requires leadership, support and resources from both those onshore and those serving onboard the vessel. Although the employer is ultimately responsible for the safety of all persons onboard the vessel, the immediate responsibility for the ship, its crew and any passengers lies with the vessel Master. All onboard also have a fundamental responsibility for their own safety.

The provision of a structured and well defined safety organisation with clear reporting links to shore based management should therefore include the shore based personnel responsible for the vessel, the Master, designated Safety Officer, appointed Safety Representatives and other members of the Safety Committee.

An example of a safety organisation is shown below; however the number and range of members will obviously depend on the marine and project crew onboard. For an Offshore Support Vessel, the Offshore Manager and Shift Supervisors would also be included.

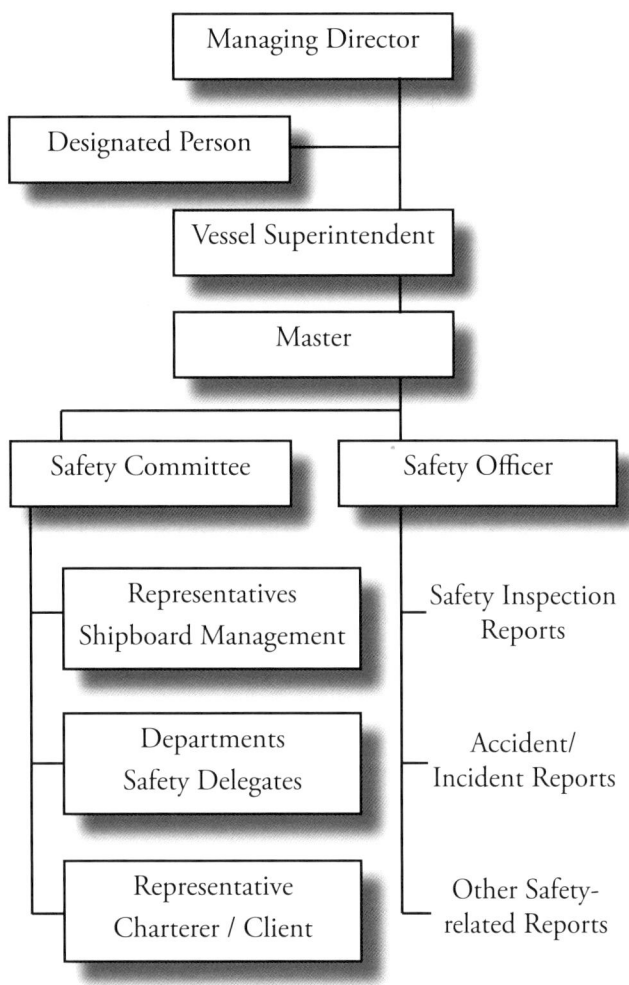

Figure 13.3 Shipboard Safety Organisation

Shore Based Management and the Master

The effective functioning of such a safety organisation will depend greatly upon the support of the shore based management and the Master. The shore based management and Master must:

- Provide the Safety Officer and Safety Representatives with access to all necessary information, documents and resources to fulfil their respective function within the safety organisation including any technical information concerning any hazards and precautions required for any machinery, substances, cargoes or operations onboard the vessel.
- Encourage the Safety Officer in his role onboard the vessel.
- Provide the Safety Officer with training if applicable to their role.
- Provide the Safety Officer and Safety Representatives with suitable time to be proactive and positive with regards to their roles in the safety organisation.
- Ensure that any representations received from the Safety Officer and Safety Representatives are considered in due time and with due care and attention.
- Ensure that any reasonable occupational health or safety suggestions or improvements are implemented whenever it is feasible to do so.

Safety Officers

On every United Kingdom ship where there are more than five workers, the company shall appoint a competent person as a Safety Officer. The Safety Officer has a key role onboard as the primary contact and advisor on all occupational health and safety issues. Due to the importance of this role, it is advisable that only someone who is keen to fulfil the function and willing to be proactive with regards to safety should be chosen to be the Safety Officer. A person appointed to the role, who is not suitable, may have a detrimental rather than positive affect. Specific Safety Officer training is available and should be provided.

It is the duty of the Safety Officer to endeavour to:

- Develop and maintain a high standard of safety consciousness among the crew and ensure that safety instructions, rules, and guidance for the ship relating to health and safety are complied with.
- Investigate, as far as is reasonably practicable, every accident involving death, major or serious injury and every dangerous occurrence onboard the vessel.
- Investigate, as far as is reasonably practicable, all potential hazards to health and safety that may exist onboard the vessel.

- Investigate, as far as is practicable, all reasonable complaints by any crew member about health and safety which relates to the vessel and its operations, unless there is any reason to believe that a complaint is of a frivolous or vexatious nature.
- Make recommendations to the Master to prevent the recurrence of an accident or to remove any potential hazard that may exist onboard the vessel.
- Ensure that health and safety inspections of each accessible part of the ship is carried out at least once every three months and more frequently if there have been substantial changes in the conditions of work.
- Make representations and where appropriate, recommendations to the Master regarding any deficiency in the vessel with regards to health and safety.
- Maintain a record of every accident involving death, major or serious injury and every dangerous occurrence onboard the vessel.
- Stop any work that is observed in progress and which may reasonably be believed to have the potential to cause a serious accident.
- Be proactive in identifying any potential hazards and for any means of preventing any future accidents.
- Provide advice on occupational health and safety for all onboard, including the Master.

Although not specified functions of the position, it is advisable that the Safety Officer should:

- Maintain a good working relationship with the Safety Representatives onboard and work with them rather than against them.
- Maintain a good working relationship with the members of the Safety Committee.

The Safety Officer has the power and authority to cease any operations onboard the vessel, which it is believed may have the potential to cause an accident.

Safety Representatives

Safety Representatives are required on all United Kingdom vessels that have six or more crew. In the case of vessels with between six and fifteen crew, one Safety Representative, as a minimum, is required to represent the officers and crew. For vessels with more than fifteen crew one Safety Representative for the officers and one for the crew are required. The Safety Representatives are not appointed by the company, but rather elected by the officers and the crew as appropriate.

Safety Representatives have powers onboard the vessel and as a member of the safety committee, however they do not have duties.

Safety Representatives and members of the Safety Committee have the power to:

- Participate, subject to the agreement of the Safety Officer, in incident investigations and inspections conducted by the Safety Officer, or conduct such investigation or inspections independently after informing the Master.
- Request that the Safety Officer conducts an investigation.
- Inspect any of the Safety Officers records including incident investigation reports and area inspections.
- Make representations to the Master or company regarding any general health and safety concerns or potential hazards and dangerous occurrences onboard the vessel.

It is advisable that the Safety Representatives should:

- Be fully conversant will all the occupational health and safety regulations which apply to the welfare of the crew and the vessel operations.
- Maintain a good working relationship with the Safety Officer, Master, Offshore Manager and Safety Committee, whilst maintaining a clear focus on representing the crew.
- Promote encouragement to the crew to be proactive with regards occupational health and safety onboard the vessel.

If the Safety Representative should find himself in a situation where his efforts are obstructed, or he is denied facilities, then he should bring the matter to the attention of the Safety Officer or to the Master through the Safety Committee.

Safety Committees and Safety Meetings

On any vessel that has at least one elected Safety Representative, the company must ensure that a Safety Committee is formed. The Safety Committee should include, as a minimum, the Master as chairman, the Safety Officer and all Safety Representatives.

The Safety Committee should be used as the forum for the Master and the appointed and elected safety officials to discuss matters relating to occupational safety onboard. Any member of the vessel crew can attend the Safety Committee meetings and for personnel with specific responsibilities for ensuring the safe and efficient operation of the vessel (Chief Engineer, Chief Officer, Offshore Manager, Shift Supervisors and Catering Officer) should also attend. However, it is also advisable that the Safety Committee should remain as compact as is practicable in order to function efficiently.

The safety meetings should be held regularly at intervals in the region of four to six weeks, however this period should be considered on a vessel by vessel basis taking into account the lengths of tours of duty of the crew and operational considerations.

The following information may be considered as the minimum discussion points at a safety meeting and may form the basis of an agenda:

- A list of attendees should be maintained including details of their position on the Safety Committee (Safety Officer, Safety Representatives, etc).
- Outstanding items from previous Safety Committee meetings should remain on the meeting minutes until all are satisfied that the item is suitably closed out.
- Details of safety area inspections completed during the period since the last Safety Committee meeting should be provided.
- Details of any accident, incident or near misses during the period since the last Safety Committee meeting should be provided.
- Details of any safety concerns raised during the meeting should be recorded, with details of the actionee and a time scale for progress to close out the item.
- Any other business relating to occupational health and safety should be recorded.

The Safety Committee meeting minutes should be displayed onboard, provided to any crew members who require them and be issued to the shore based management for the vessel.

Safety Area Inspections

The appointed Safety Officer is required to inspect each accessible part of the vessel at least every three months and to maintain records of the inspections which should include any recommendations and actions taken with regards to occupational health and safety. It is a general guide that the vessel should be divided into twelve areas; therefore one inspection each week will fulfil the three monthly inspection requirements. This will of course depend on the type and size of the vessel and should, if at all possible, be divided into recognisable areas. A typical twelve area inspection rota for a Dive Support Vessel could include the items shown in figure 13.4.

Prior to an area inspection, the Safety Officer should review the previous inspection report for that area and note any recommendations, actions or areas of concern that were raised previously. Such issues should be considered during the inspection.

The items that should be considered during any area inspection will change dependent on the area. Specific checks will have to be considered when inspecting areas

No.	Area
1	Monkey Island
2	Helideck
3	Exterior accommodation
4	Interior accommodation, inc. Galley
5	Engine room
6	Forward thruster spaces
7	Aft thruster spaces
8	Dive system handling area
9	Dive system saturation chamber area
10	ROV launch and recovery areas
11	Main after deck
12	Cranes

Figure 13.4 Dive Support Vessel - Safety Inspection Areas

such as the galley, particularly with regard to health and hygiene and therefore the purpose, environment and operations that are related to the area concerned should be considered during the inspection.

Accident and Incident Reporting

The Merchant Shipping (Accident Reporting and Investigation) Regulations 2005 (as amended) have been in force since 18th April 2005.

The regulations apply to all United Kingdom vessels and all vessels operating within United Kingdom waters with certain exceptions. The information provided below refers to a vessel registered in the United Kingdom.

For the purposes of the regulations, an accident means any occurrence onboard a ship or involving a ship whereby:

- There is a loss of life or major injury to any person onboard or any person is lost or falls overboard from the vessel or one of its boats.
- The vessel causes any loss of life, major injury or damage.
- The vessel is lost, presumed lost or abandoned.
- The vessel is materially damaged by fire, explosion, weather or other cause.
- The vessel grounds, is in a collision, is disabled or causes significant harm to the environment.
- There is a collapse or bursting of any pressure vessel, pipeline or valve.
- There is a collapse or failure of any lifting equipment, access equipment, hatch cover, staging or boatswains chair or any associated load bearing parts.
- There is a collapse of cargo, unintended movement of cargo or ballast sufficient to cause a list, or loss of cargo overboard.

- There is an escape of any harmful substance or agent.

If any of these, or any specific occurrence as detailed in the Regulations, occur onboard, the incident must be reported. The Regulations provide full details of the reporting requirements, including:

- Definitions of relevant terms, including detailed definitions of the terms such as 'major injury'.
- The information to be included in any report.
- The circumstances which would require an incident investigation by the Chief Inspector.
- Details of the conduct of an investigation by the Chief Inspector.
- Details of the records that the inspector can access as a result of such an investigation.
- Penalties that may be imposed by the Chief Inspector.

Accident and Incident Investigations

The Safety Officer has a statutory duty to investigate any accidents or incidents onboard and should do so at the soonest opportunity following the incident. The following summary may provide guidance on the steps that the Safety Officer should take in such a situation.

Investigation of the Incident Area

- The immediate priority must be to attend to any injured parties at the scene and secure the integrity of the area as soon as possible. Securing the scene may prevent further injury or incident.
- Once sufficient assistance has arrived and any injured parties are being attended to, the Safety Officer should concentrate on establishing the facts concerning the incident. If at all possible, the Safety Officer should leave the rescue operation and first aid to other qualified and experienced personnel in order to concentrate on gaining as much information and an immediate overview of the incident as is possible.
- All persons, including any injured parties, witnesses and other persons connected with the incident should be identified and their details collected for future reference during the investigation.
- The position of any injured parties should be marked and the current condition of the area, equipment and systems in use should be noted. Photographic evidence is essential.
- Following on from the preliminary inspection, a more thorough inspection of the area must be conducted, ensuring that any changes that may have occurred since the time of the incident are considered such as environmental factors (lighting, rain, ice), any equipment that may have moved or any changes in the area structure (accesses closed etc). Any notable hazards that may have been a contributory factor in the incident should also be noted and photographed.

Figure 13.5   Skandi Navica — Pipe Lay Vessel

Witness Statements

It is essential to conduct witness interviews as soon as possible in order that the witnesses will have the details of the incident fresh in their minds. It is also not advantageous for witnesses to have discussed the incident prior to the interview where they may be swayed in their opinion of what happened by others who were present or involved.

In addition, any persons who may have not been present at the time of the incident, but may have been involved in earlier discussions concerning the operation that was being performed or who may responsibility for the work or area of the operation should also be interviewed. Their input may have significant relevance to the cause of the incident.

Any interview of a direct witness or a secondary person should be conducted in an informal atmosphere in order to try and relax the witness and put them at their ease and should follow a structured approach which should be followed as closely as possible for all interviews relating to the same incident.

- The purpose of the interview should be explained.
- Key background details of the interviewee should be ascertained.
- The witness should be requested to relate the incident as accurately as possible and from start to finish with little if no interruption. The witness should be requested to concentrate on the facts and avoid expressing opinions. Any inconsistencies between this version of the events and any others provided

and also any details that may require to be clarified should be noted and should be checked after the initial complete run through of events.

- Once clarification and further questioning has been completed, the interviewer should repeat the details of the statement to the witness in order to check that they have been accurately recorded.
- If satisfied with the accuracy of the statement, the witness should sign and date the document. However, if the witness changes his or her mind concerning any details during the preparation of the witness statement, then the statement should be retained and annotated by the Safety Officer as such.
- The witness statement(s) should be maintained with the incident report and any the relevant details and documents, by the Safety Officer.

Inductions

Essential to ensuring a safe working environment onboard for any new crew members, third party personnel or visitors, is the provision of a suitable induction to the vessel and essential information concerning safety onboard.

The specific information and areas that should be covered within a safety induction and tour will depend on the position of the person being inducted. For example, a crew member may have access to the engine room and therefore certain aspects of engine room operations and safety equipment must be explained. However, a third party contractor who will only visit the bridge will obviously require a less expansive tour and induction.

However, irrespective of position, rank or area of work, the following basic areas should be covered for a general vessel induction, with particular relevance to an Offshore Support Vessel:

- The general alarm and fire alarms details and any mustering points onboard the vessel or on the quayside during port calls are the most important information for any induction and should be provided to all personnel, including new crew members, sub-contractors and visitors.
- Emergency exits and escape routes, particularly in the areas where the person may be working and any relevant safety equipment, such as fire alarm activation points and fire extinguishers, should be identified. The location of safety plans should be indicated; where more extensive details of the life saving and fire fighting equipment onboard can be obtained.
- For permanent marine crew, a process of familiarisation will include the location and use of all life saving and fire fighting equipment including life rafts, fixed fire fighting systems, ventilation flaps and fuel and air stops.
- The personal protective equipment requirements onboard the vessel will be applicable to marine crew, project personnel and third party sub contractors. The requirements for different areas should therefore be conveyed to all crew and visitors.
- The shipboard Safety Management System may only be applicable to certain new crew members; however aspects of the system such as company policies and procedures may be applicable to all. For example, a sub contractor working on deck will need to be made aware of the designated smoking areas onboard the vessel, drug and alcohol policy and most importantly for third party personnel, the permit to work system and isolations process. Marine crew will of course require to be familiarised extensively with the requirements of the Safety Management System.
- The location of the hospital and first aid stations and the responsible person onboard for medical issues should be identified. Emergency contact numbers whilst in port should be identified and posted.
- The Safety Officer and Safety Representatives should be identified and a single point authority (SPA) identified for any sub contractors working onboard. Therefore any operations being conducted in the engine room, for example, should be under the control of the Chief Engineer. Sub contractors should therefore be introduced to the relevant authority for liaison during the task at hand.
- The location and use of onboard oil pollution prevention equipment may be applicable to marine crew, project personnel and sub contractors.
- If the individual will be required to work in areas where powered watertight doors are located, they should be made aware of the associated dangers and if necessary, instructed in their operation, particularly in an emergency.
- Any onboard site specific requirements such as general housekeeping and waste segregation are essential in order to maintain the vessels routine and to avoid the production of hazards due to a failure in basic housekeeping.
- Particular hazards associated with Offshore Support Vessels must be conveyed to any visitors, sub contractors and new crew members, especially if they will be onboard during any offshore operations. The danger areas, particularly on the after deck, must be highlighted.
- Any specific induction requirements for Dive, ROV, Construction or Pipe Lay Vessels and their respective

equipment will be required for specific crew and project personnel.

All personnel who are subject to a shipboard induction should be required to sign off affirming that they have been provided with a suitable induction and that they have understood the requirements that have been explained to them.

Figure 13.6 Toisa Polaris — Dive Support Vessel

Safe Access and Slips, Trips and Falls

Along with manual handling, slips, trips and falls are amongst the most common causes of injuries onboard ship. Such injuries can occur in almost any area onboard the vessel from the gangway to the galley and from the main deck to the pilot ladder. This section therefore provides a general overview of the risks associated with safe access to the vessel for both crew members via the gangway and for pilots via the pilot ladder and also for general onboard access arrangements which includes the maintenance of free and unobstructed walkways and working decks, vertical ladders and stairways.

Gangway and Accommodation Ladder Access

The Merchant Shipping (Means of Access) Regulations 1988 and the Code of Safe Working Practices for Merchant Seamen provide detailed information and guidance concerning the provision of a safe means of access from the vessel and quayside or any other ship or pontoon which may be moored alongside.

This means of access must be positioned promptly and remain in place while the vessel is secured, be properly rigged and secured, be adjusted from time to time as environmental conditions, such as the tide, dictate. The means of access should be adequately illuminated and be provided with a suitable safety net which must extend the entire length of the access ladder and must provide protection at both the upper and lower extremities of the access way.

Gangways must be carried on ships of 30 metres in length or over and accommodation ladders must be carried on ships of 120 metres in length or over.

Although gangways and accommodation ladders differ in nature with accommodation ladders being permanently fixed to the vessel and gangways being a moveable and temporary arrangement, the safety considerations required to be observed when they are in position can be considered to be fundamentally similar.

- Gangways and accommodations ladders must be clearly marked with the manufacturer's name, the model number, the maximum designed angle of use and the maximum safe loading both by numbers of persons and by total weight.
- An accommodation ladder should be designed and located such that it rests firmly against the ship's side.
- The angle of slope should not be greater than 55 degrees to the horizontal. Steep angles of ascent and descent must be avoided.
- Treads and steps should provide a safe foothold over the range of angles for which the ladder is designed to be used within.
- Suitable fencing (preferably rigid handrails) must be provided along the entire extent of the gangway or accommodation ladder, excepting that the fencing at the bottom platform may allow access from the outboard side.
- The lower and upper access platforms, if fitted, should be horizontal and any intermediate platforms should be self levelling.
- Any slip hazards caused by ice, snow or any contaminant should be removed immediately or treated appropriately.
- Suitable safety equipment such as a life buoy with self-activating light and buoyant safety line should be provided ready for use at the point of access aboard the ship.

Figure 13.7a Gangway Access

Figure 13.7b   Accommodation Ladder Access

Safety nets should extend the entire length of the ladder and provide fall protection at both the upper and lower access platforms. Figure 13.7a shows an inadequate net which provides only limited protection. Treads should be suitable for the range of angles of use the ladder is designed for such as these provided by specialist manufacturer Scotgrip.

Although the majority of existing accommodation ladders and gangways are made of aluminium, which in itself can provide a hazardous slippery surface, products such as the treads provided by Scotgrip can improve the access arrangements greatly. These treads provide a high traction surface and can be fitted onto existing ladders mechanically or adhesively.

General Onboard Access

The provision of safe access ways onboard the vessel can be mitigated by good housekeeping practices and by routine inspection of all access walkways and working decks, vertical ladders and stairways. Maintaining free and unobstructed passage for all crew members and the provision of an anti-slip and contaminant free deck can significantly reduce the possibility of slips, trips and falls.

Walkways and Working Decks

The working decks and areas onboard ship can be constructed from a variety of materials including flat steel, open grating, chequered plate, aluminium and timber. These surfaces are acceptable when dry, however due to the working environment, that is rarely the case, and these deck surfaces can be extremely dangerous when wet. To counter such hazards there are a number of options available to help provide a more stable deck surface, including anti-slip paint applications and deck coating products. Deck tiles are available which can be easily fitted onto the majority of surfaces and provide a tough and durable, high traction surface capable of preventing slips, trips and falls and also providing an extended lifespan of the underlying deck surface.

Similarly, obstructions on the deck, such as pipe work and cabling is the main permanent hazard onboard any type of vessel. Generally it is almost impossible to negate the need for deck pipe work to be exposed; however it is possible to provide suitable shielding using products such as the Scotgrip pipe and cable bridges as shown in figure 13.8.

Figure 13.8   Pipe and Cable Bridges

Pipe and cable bridges can be utilised as temporary installations, for example to provide a clear walkway and protection for pipe work and hoses during a dry docking. Alternatively, they can also be used as permanent fixtures to conceal fixed pipe work.

Vertical Ladders and External Stairways

Vertical ladders are no different to horizontal working areas and can provide a slip hazard. This can be due to the inherent smoothness of the construction material and also due to environmental conditions. However, treads are available for vertical ladders that can be fitted on square bar or rounded bar rungs. These rungs are secured in place by bonding.

*Figure 13.9
Round Bar Rung Covers*

The leading edge of a step is the most dangerous part, especially if it has been damaged, worn smooth or rounded by constant use. The rear edge of a step can also be a concern if damaged. The provision of kicker treads such as those shown in the adjacent photographs can not only reduce the slip hazard, but also provide protection and increase the life span of the step structure.

Figure 13.10 Kicker Treads on External Stirways

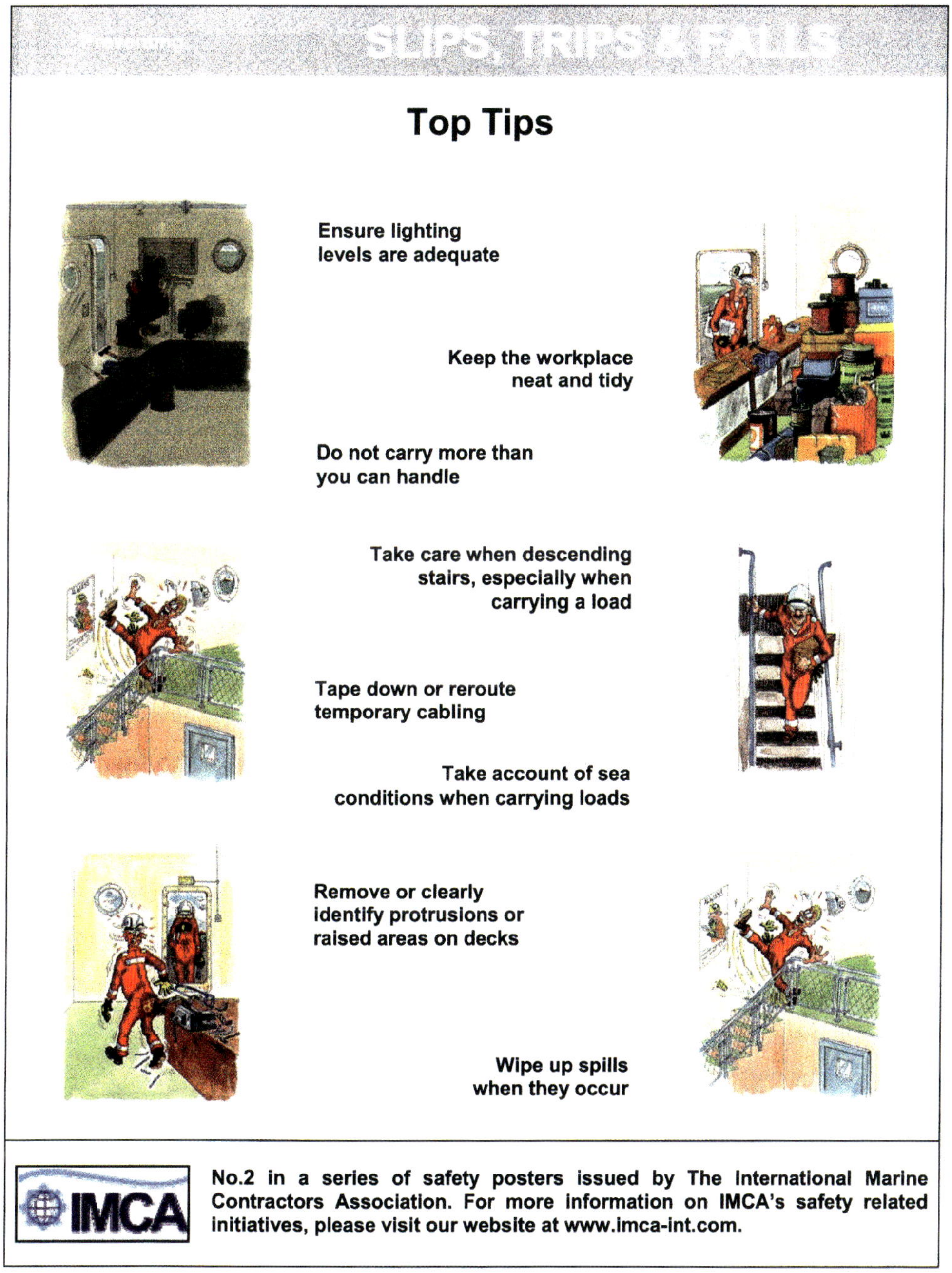

Figure 13.11 IMCA Slips, Trips and Falls Poster

Pilot Access Arrangements

Pilot boarding is a major safety hazard on any vessel. The low freeboard associated with Offshore Support Vessels may suggest that these hazards are not present during pilot boarding and disembarking, however low freeboards can cause their own dangers and therefore some general guidance on pilot ladder and pilot boarding and disembarking arrangements are provided here.

Figure 13.12 Pilot Boarding Operations

Safety of Life at Sea Chapter V: Safety of Navigation

Chapter V of SOLAS and the Merchant Shipping (Pilot Ladders and Hoists) Regulations 1999 detail the requirements for the provision of a safe access for pilot boarding and disembarkation. Further details are contained within the International Maritime Organisation Resolutions A.426 (XI) - Arrangements for Embarking and Disembarking Pilots in Very Large Ships, A.275 (VIII) - Standards for Mechanical Pilot Hoists, A.667 (16) - Pilot Transfer Arrangements and A. 889 (21) - Recommendation on Pilot Transfer Arrangements.

In addition, the Maritime and Coastguard Agency provide practical guidance on pilot boarding operations within the Code of Safe Working Practices for Merchant Seamen.

What all of the above legislation and guidance has in common is the need to provide a safe means of access for the pilot when boarding a vessel in the open sea, ensuring that all the equipment utilised is in good condition and used in the correct manner. The provision of a stable and sheltered platform (the vessel) for the equipment is of course essential. The legislation regarding pilot boarding arrangements is extensive, however in practical terms, the following should be considered.

- The pilot ladder must be in good condition and therefore storage before and after use is an important factor. Pilot ladders should be stored clear of any contamination sources, including paint, direct sunlight and chemicals. The regular maintenance and inspection of the pilot ladder should be incorporated into the planned maintenance schedule.

- The pilot ladder and associated equipment should not be used for any other operation.

- Prior to rigging, the ladder must be inspected to ensure that the structure and all component parts are free from grease, oil or other forms of contaminant. Any such contamination may pose a slip hazard or provide a long term hazard to the integrity of the pilot ladder.

- The point of access to the vessel at the top of the pilot ladder should be suitably free of any possible obstructions or hazards (such as overhangs or vessel equipment) which may cause injury to a pilot accessing the vessel.

- The gateway at the top of the pilot ladder to the vessel's deck should provide a clear access to the vessel. However, where a pilot ladder is rigged over a bulwark or handrail, a suitable means of access must be provided from the top of the pilot ladder to the deck (bulwark ladder).

- Stanchions, when fitted, should be secured in such a manner that they cannot be inadvertently removed by accident and should provide a safe and rigid handhold at the point of access to the vessel deck.

- Lighting (permanent or temporary) should be suitable to ensure good illumination of the ladder, point of access from the pilot launch and the point of access from the pilot ladder onto the vessel. However, lighting should not be rigged in such a manner that there is the potential to dazzle the pilot or pilot launch crew.

- The rigging of the pilot ladder and the embarkation and disembarkation of the pilot should be supervised by a responsible officer, who should arrange for the safe escort of the pilot to and from the bridge. The officer should ensure that the ladder and man ropes are securely made fast prior to the arrival of the pilot launch.
- There should be a suitable means of communication available between the pilot embarkation point and the bridge.
- The pilot ladder should rest firmly against the ship's side and be clear of any discharges.
- Two man-ropes of not less than 28 mm in diameter, properly secured to the vessel, a lifebuoy equipped with a self-igniting light and a heaving line should be readily available for immediate use at the pilot ladder.
- The rigging of the pilot ladder should take into consideration the prevailing weather conditions, the draft and trim of the vessel and the pilot boarding location.

Risk Assessments and Hazard Identification

The Health and Safety at Work Regulations require all employers to evaluate all risks to which their employees may be exposed. One of the most valuable tools in the assessment of risk onboard any vessel is the risk assessment. A risk assessment can be defined as a careful and considered examination of an intended operation and the identification of any associated hazards. The risk assessment methodology should be used to mitigate against any of the identified hazards to remove all practicable risks from the operation.

Risk assessments should be conducted onboard prior to commencement of work on any task for which no valid risk assessment has been conducted. In any case, if a previous risk assessment is in place for a particular operation, then this assessment should be reviewed prior to the work being carried out with particular attention to any changes in the proposed operation. A risk assessment should include:

- An assessment of the work activity planned.
- Identification of the hazards associated with the planned operations.
- A review of the controls in place as part of the vessels management system.
- An evaluation of the personnel at risk.
- A review and evaluation of the potential for harm to the personnel involved.
- A review and evaluation of the potential severity of any possible harm.
- An assessment of the 'risk levels' involved in the operation.
- Any actions to be taken as a result of the risk assessment in order to reduce or eliminate any of the identified hazards.

'Permit to Work' Systems

Permit to Work Systems have been used onboard ships for many years now and all mariners will be familiar with the basic format and use of permit systems. Each shipping company will have specific requirements for their own permit to work system; however the purpose and principles remain generic.

Hazard Severity Category	Descriptive Word	POTENTIAL CONSEQUENCES			PROBABILITY RATING					
		Personal Illness / Injury	Equipment Loss	Environmental	A	B	C	D	E	F
I	Catastrophic	Fatal or permanent disabling injury or illness	> $1,000,000	Potentially harms or adversely affects the general public	1	1	1	2	3	3
II	Critical	Severe injury or illness	$200,000 to $1,000,000	Potentially harms or adversely affects trained employees and the environment at our facility	1	1	2	3	3	3
III	Marginal	Minor injury or illness	$10,000 to $200,000	Presents limited harm to the environment and requires general expertise and resources for correction	2	2	3	3	3	3
IV	Negligible	No injury or illness	<$10,000	Presents limited harm to the environment and requires minor corrective actions	3	3	3	3	3	3

A: Frequent   B: Reasonably probable   C: Occasional   D: Remote   E: Extremely improbable   F: Impossible

Figure 13.13   Risk Analysis Matrix

The Purpose of the 'Permit to Work' System

The main purpose of any Permit to Work System is to create a safe system of work. In order to provide such a safe system, the permit to work should:

- Ensure the safety of the crew and integrity of the plant, machinery and vessel.
- Provide the means for ensuring good co-ordination between work teams onboard the vessel, cross department operations and external operations.
- Ensure that any plant or equipment is safe, prior to commencement of any maintenance or task.
- Assess all the risks associated with the intended task, working environment and external sources and ensure that they have been systematically identified and effectively controlled prior to commencement of the task.
- Ensure compliance with any international, local, statutory or contractual requirements.

The Principles of the 'Permit to Work' System

In order to ensure that a safe system of work is provided, the Permit to Work must:

- Ensure that performing authorities are given clear and well defined directions.
- Be clear and unambiguous, stating the task to be done.
- Clearly identify and describe the plant and equipment involved.
- Provide a statement of the extent to which the plant is made safe.
- Provide a warning of any possible remaining hazards and control such residual hazards.
- Provide confirmation that any isolation (mechanical or electrical) is adequate and ensure that it remains secure.
- Provide acceptance of the task and agreement to abide by the precautions specified.
- Provide notification that the task is complete or incomplete.
- Provide acceptance that the task is complete or incomplete.
- Be monitored and audited at regular intervals.

Entry into Enclosed Spaces

Introduction

Entry into enclosed spaces has always been one of the most hazardous areas of work onboard ships and has, over the years, claimed many lives. In many of these incidents failure to follow basic precautions before entry and a lack of appreciation of the risks involved has sadly led to multiple deaths. In some cases, personnel attempting to rescue colleagues who have collapsed in enclosed spaces have often died during their attempts to save the lives of their fellow crewmembers. By being aware of the potential dangers and hazards associated with enclosed spaces and by taking satisfactory precautions before and during any enclosed space entry operation, such deaths are avoidable.

What is an Enclosed Space?

The main space onboard a ship that most people would associate with the term 'enclosed' is undoubtedly a fuel or cargo tank. However, the number and variety of spaces onboard that can be described as such and which pose potential dangers to personnel are extensive.

The following can be considered 'enclosed spaces':

- Fuel Tanks
- Ballast Tanks
- Fresh Water Tanks
- Cargo Tanks
- Sewage Plant Tanks
- Stability Tanks
- Chain Lockers
- Void Spaces / Cofferdams
- Pump Rooms
- Duct Keels
- Equipment Spaces (for log transducers etc)
- Main Engine Crankcases
- Scavenging Air Spaces

In summary, the term 'enclosed space' can be defined as any location that has limited openings for entry or egress and is not intended for continuous human occupancy.

What are the Dangers?

Dependent on the space that is being considered and the type of substance that may have been stored in the space, the potential dangers that may exist can include:

- Oxygen Deficiency
- Toxicity
- Flammability
- Other hazards such as contact with chemicals and corrosive substances.

Assessment and Control

Entry into an enclosed space should be conducted in accordance with the Permit to Work System and should be thoroughly risk assessed prior to commencement of the proposed task.

A competent person should be assigned to assess the proposed task and the space to be entered. This person should be in charge for the duration of the work and should consider the following factors during the assessment of the task:

- Hazards - The assessment of the hazards associated with the tank should include a review of the substances that have been carried in the tank. For example, an empty tank may be deficient in oxygen due to the oxygen content combining with steel in the rusting process.
- Isolation & Ventilation - the space should be isolated from potential ingress of dangerous substances. For example, fuel tank valves should be locked off. Thorough ventilation of any space by natural or mechanical means must be carried out.
- Atmosphere - In accordance with the hazard assessment detailed above, the space should be tested for oxygen content, flammability and / or toxicity.

Prior to Entering an Enclosed Space

The risk assessment and Permit to Work should determine the practical requirements prior to entry, however the following is advised:

- The access point should be well illuminated. Any lights used inside the tank must be approved for use in a flammable atmosphere, if appropriate.
- The means of access (ladder or walkway) should be in good condition.
- Rescue and resuscitation equipment should be maintained at the access point including Breathing Apparatus and a recovery stretcher. It should be confirmed that such equipment can easily be transferred into the space.
- One person should be assigned to remain at the access point at all times.
- Direct and continuous communication must be available between the access point and the Officer of the Watch.
- A system of communication between the person(s) entering the enclosed space and the person remaining at the access point should be agreed.
- It may be considered prudent that anyone entering the space wear a rescue harness to facilitate recover in an emergency situation.

Continuous Monitoring

- Ventilation - the space should continue to be ventilated during the task.
- Atmosphere - the atmosphere should continue to be tested regularly throughout the task.
- Hazards - any changes in the potential hazards that may affect the operation should result in the work being suspended and all personnel being removed from the enclosed space. Work should not recommence until a review of the task is conducted.
- Emergencies - in the event of an emergency, the Officer of the Watch should be contacted and general alarm sounded immediately. Under no circumstances must anyone enter the tank until back-up personnel are available and breathing apparatus must be worn.

United States Coast Guard - Advice on Reducing the Risks

Many organisations have issued guidance on procedures for entry into enclosed spaces, including the MCA Code of Safe Working Practices, IMO Recommendations for Entering Enclosed Spaces Aboard Ships and the International Association of Classification Societies (IACS). Summarised below is the advice issued by the United States Coast Guard.

- All vessels complying with the International Safety Management Code should have a specific plan for entering confined spaces outlined within their Safety Management System.
- The confined space entry procedures should include and identify various types of shipboard spaces which should be treated as confined spaces.
- Crew safety meetings should address the identification of confined spaces and provide instruction on confined space entry procedures.
- Individual crewmembers that work in confined spaces should review existing entry procedures and requirements regularly.
- Evacuations and rescues must be well planned and rescuers fully protected.

Mooring Operations

Mooring operations have always been a high risk activity and every year many seafarers and rope men are seriously or fatally injured as a result of mooring accidents. Irrespective of experience, it is therefore beneficial to re-assess the dangers that may be present regarding mooring equipment and mooring operations. The following text highlights some of these dangers.

Installation of Mooring Equipment

The importance of the type, rating, systems and positioning of the mooring equipment onboard is fundamental to safe mooring operations.

- Mooring winches and windlasses should be positioned in such a manner that there is no requirement for any crew member to be positioned within the bight of a rope lead.

- Mooring winches and windlasses should be fitted with an alarm system to give warning of undue strains, either by stalling or by 'walking-back' at a specific load.
- Mooring rollers, fairleads and bitts should be properly designed to ensure their positioning is advantageous to safe mooring operations and suitable for the expected loads for the particular vessel size.
- Mooring ropes should be sufficient to ensure that the vessel can be safely moored. However they should also be commensurate with the mooring equipment fitted. Mooring ropes with a higher strength rating than the mooring system can result in failure of the winches with potentially fatal results.
- Decks should have anti-slip surfaces provided by fixed treads or anti-slip paint coating, and the whole area should be adequately lit for periods of darkness.

Planned Maintenance and Repairs

Planned maintenance for all mooring equipment should be in line with manufacturer's instructions augmented with any vessel specific considerations and any specific requirements due to weather conditions (hot / cold).

To complement the planned maintenance program, regular inspection for damage, defects, wear or corrosion should be completed. Any signs of stress or fracture may highlight a potential hazard which would not normally be noted unless the item of equipment was under stress.

In addition to the permanently installed mooring equipment, the mooring ropes themselves should be maintained in good condition and should be inspected regularly, including before use, for any damage or defect. Storage of the mooring ropes should be such that they are free of contamination from chemicals or sunlight. Replacement ropes should be used if there is any doubt surrounding the condition of the ropes.

Safety Considerations during Mooring

- An overview of the mooring station is essential to understanding the potential 'snap-back' zones that may be present. An example of such a diagram is given in the Code of Safe Working Practices for Merchant Seamen.
- The danger areas and rope or wire leads and bights are obviously linked to the design and location of the mooring equipment and leads. However, a review of whether the safest mooring arrangement is being used considering the restraints the design of the equipment imposes, may help to identify potential risks and any alternative arrangements that may be possible.
- Only personnel directly involved with mooring operations should have access to the mooring stations during operations. In addition, access should also be limited when moored alongside. The mooring stations can be considered high risk areas when secured due to the changing tide levels, vessel draft and other vessel movements. Due to these factors the strain on mooring equipment can be continually changing causing hazards, even when the vessel is safely moored alongside.
- All personnel assigned to a mooring station should be provided with appropriate personal protective equipment, including eye protection when operating windlasses.
- Mooring ropes and wires which are stowed on reels should not be used directly from their stowage position. They should be run off and flaked on deck in a clear and safe manner, to ensure that sufficient slack will be available during deployment. This will avoid excess loads being applied due to snagging.
- Mooring ropes and wires should not be deployed through the same fairlead or secured on the same bitts. Such practices are likely to lead to damage to the ropes or wires and snagging.
- In most circumstances three turns on the drum end are sufficient when mooring.
- Communications between the bridge and mooring stations and between the personnel at each station are critical. If this should involve the use of portable radios, then the vessel name should be clearly identified to prevent misinterpretation.

Guidance on mooring operations is provided by a number of authorities and in a number of publications.

Figure 13.14   Mooring ropes should be checked regularly

Figure 13.15
Mooring ropes should be stored clear of any contaminants

However, some of the most practical and common sense advice is provided by the Code of Safe Working Practices for Merchant Seamen and the Institute's book *The Mariner's Guide to Mooring*.

Manual Handling

Manual handling has long been an accepted part of many operations onboard Offshore Support Vessels and accounts for a high percentage of injuries.

The term manual handling is not restricted to lifting or carrying objects, but is also used to describe pushing, lowering, pulling, holding or restraining. Therefore manual handling does not always include the lifting of an object, but can also be used to describe:

- Tasks that involve sudden or fast movements.
- Tasks that involve bending, reaching or twisting of the body.
- Tasks that require the individual to remain in a static position or posture for a long time.
- Tasks that are repetitious.
- Tasks that require heavy loads to be lifted and carried or that require force or effort by the individual.

It is accepted that manual handling onboard an Offshore Support Vessel poses many additional hazards, in particular the sudden and unexpected movement of the working environment. However, the following factors should be considered for any manual handling task:

- The ergonomics of the workplace should allow the individual to maintain a good posture.
- Repetitious, fast and twisting actions should be avoided if at all possible.
- Heavy or awkward loads should not be stored above shoulder or below mid thigh height to allow the individual to maintain a good stance when lifting or lowering the load.
- The weight of the load should be ascertained and the use of lifting appliances should be considered or two personnel instead of a single individual.
- Lighting levels should be adequate and the floor surface should be free of obstructions and spillages.
- Appropriate personal protective equipment should be worn dependent on the task being performed and the environmental conditions.
- The properties of the load should be considered, including the shape, contents, centre of gravity and independent movement of the object.
- The individual involved in the operation should be aware of the risks and have been provided with suitable manual handling training.

Prior to a manual handling task, the risk associated with the task should be controlled. The most appropriate method to reduce the risk of manual handling injuries is to re-design the task or adapt the environmental conditions to make the task safer for the individual. This may require the object to be modified, the work area to be modified, complete the task utilising mechanical lifting appliances, levers or by team lifting.

Lifting Operations and Lifting Equipment

The Health and Safety Executive (HSE) 'Lifting Operations and Lifting Equipment Regulations 1998' came into force on 5th December 2002.

The Merchant Shipping and Fishing Vessels (Lifting Operations and Lifting Equipment) Regulations 2006, came into force as of 24th November 2006. These regulations apply to all United Kingdom vessels and all vessels within United Kingdom waters.

The Regulations provide details of the:

- Duties of the Employer with regards to lifting operations and lifting equipment.
- Strength, stability, positioning, installation and suitability requirements of lifting equipment for the intended purpose and scope of work.
- Marking and certification requirements of lifting equipment.
- Organisation requirements and controls for lifting operations.
- Maintenance requirements for lifting equipment.
- Testing, thorough examination and inspection requirements of lifting equipment.
- Training and competence requirements for personnel involved in lifting operations.
- Records to be maintained.

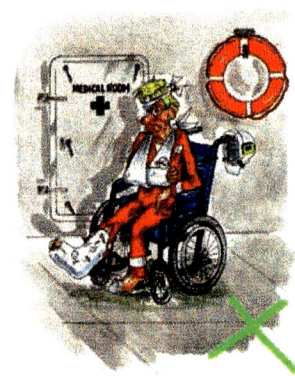

Figure 13.16   IMCA Manual Handling Poster

Figure 13.17   Toisa Perseus — Pipe Lay Vessel (prior to installation of Pipe Lay System)

Provision and Use of Work Equipment

The Provision and Use of Work Equipment Regulations (PUWER 98) applies to offshore installations, wells, pipelines and pipeline works, and to connected activities within the territorial waters of the United Kingdom or in designated areas of the United Kingdom Continental Shelf.

For merchant vessels, The Merchant Shipping and Fishing Vessels (Provision and Use of Work Equipment) Regulations 2006 came into force from 24th November 2006. These regulations apply to all United Kingdom vessels and all vessels within United Kingdom waters.

The Regulations provide details of the:

- Duties of the Employer with regards to the provision and safe use of work equipment.
- Suitability requirements of work equipment for the intended purpose, taking into consideration the working environment and ergonomic principles.
- Maintenance requirements for work equipment.
- Inspection and certification requirements for work equipment.
- Instructions and information that must be provided to the user of any work equipment.
- Training that must be supplied for the use of work equipment.
- Assessments that must be undertaken to ensure that all suitable guards are in place for dangerous parts of work equipment, isolation requirements for electrical equipment and protection against specified hazards.
- Controls, emergency controls and stops requirements for work equipment.
- Markings and warnings required for work equipment.

Working at Height

The Work at Height Regulations 2005 came into effect on 6 April 2005. The Regulations apply to all work at height where there is a risk of a fall liable to cause personal injury and have been adopted on many Offshore Support Vessels. The Regulations place duties on employers and any person that controls the work of others. These duties include ensuring that:

- All work at height is properly planned and organised and those involved in work at height are competent.
- The risks from work at height are assessed and appropriate work equipment is selected and used. The risks from fragile surfaces should be properly controlled.
- The equipment for work at height is properly inspected and maintained.
- Collective measures to prevent falls (such as guardrails and working platforms) are in place.

To ensure that these duties are met, the duty holders must:

- Avoid work at height where possible.
- Use work equipment or other measures to prevent falls where they cannot avoid working at height.
- Where they cannot eliminate the risk of a fall, use work equipment or other measures to minimise the distance and consequences of a fall should one occur.

Working at height can include a variety of activities that may affect any number of personnel whether they are marine deck crew, project crew, marine engineers or deck officers. The situations which can be encountered are many and for example could include planned

Information

Working at height is defined as working at a place from which a person could be injured by falling from it, regardless of whether it is above, at or below ground level.

maintenance on a radar scanner on the main mast, loading pipe or overhauling a main engine where the engineer has to work on the top of the main engine. These are areas where it is very obvious that working at height is a concern, however it should be remembered that there are a number of activities onboard where other dangers may be focused upon more intently, but working at height dangers still exist.

Work inside a ballast tank, for example, may pose considerable exposure to falls and working at height precautions should be considered. Activities such as deploying the gangway may similarly include some exposure to falls.

Risks involved with working at height can therefore be present in situations where it is not considered the main concern. However, failure to define the work as 'at height' would be a serious error.

The following general precautions are applicable:

- Working at height should be avoided if at all possible.
- The movement of the vessel at sea or mooring alongside should be considered.
- All personnel working aloft should wear a safety harness with lifeline or other fall arresting device.
- Where work is to be carried out at height and over the side, buoyancy aids should be worn and a lifebuoy and line should be available for immediate use.
- Other than in emergency situations personnel should not work over side when the vessel is underway.
- When working in the vicinity of any electrical equipment (such as radar scanners), the electrical equipment must be isolated at source and the permit to work and isolation procedures should be followed.
- Any tools or equipment used when working aloft must be secured and care taken to avoid dropped object hazards. Warning signs and barriers should be used in areas below any such work.
- Portable ladders should only be used where no safer means of access is reasonably practicable.
- Portable ladders should be pitched between 60 and 75 degrees from the horizontal, on a firm and stable base. The ladder should be properly secured.
- Portable ladders should be secured or tended whilst in use.

Chapter 14

ENVIRONMENTAL MANAGEMENT

General Introduction

The following section provides general guidance on environmental management with particular emphasis on vessel emissions and the disposal of waste materials. For the purposes of this text, it is assumed that all vessels will comply with the requirements of MARPOL whether ratified or not and, in addition, with the UK Environmental Protection Act 1990. This information is therefore primarily for vessels operating in the UK Sector of the North Sea, however it may be used as a guide to best practice for vessels operating worldwide.

For the purposes of this guidance, emissions and waste will be examined as follows:

- MARPOL Annex I – Regulations for the Prevention of Pollution by Oil.
- MARPOL Annex IV – Regulations for the Prevention of Pollution by Sewage from Ships.
- MARPOL Annex V – Regulations for the Prevention of Pollution by Garbage from Ships.
- MARPOL Annex VI – Regulations for the Prevention of Air Pollution from Ships.
- Record Keeping Requirements (MARPOL).
- Garbage Management (Environmental Protection Act 1990).
- Record Keeping Requirements (Environmental Protection Act 1990).

In addition, a summary of the International Standards Organisation (ISO) 14001 Environmental Management Standard is provided. Although not mandatory for any vessel, many offshore operators are certified in accordance with the requirements of ISO 14001 or may be expected to do so in the future. A basic understanding of the system may therefore be beneficial.

MARPOL Annex I – Regulations for the Prevention of Pollution By Oil

Annex I of MARPOL 73/78 - Regulations for the Prevention of Pollution by Oil applies to all ships.

Control of Discharge of Oil

For a vessel of 400 gross tonnes and above, other than an oil tanker, any discharge into the sea of oil or oily mixtures shall be prohibited except when all of the following provisions have been met:

- The vessel is not within a special area as defined in Annex I of MARPOL 73/78.
- The vessel is proceeding en route and making way through the water.
- The oil content of the discharge without dilution does not exceed 15 ppm.
- The vessel has in operation oil filtering equipment (oily water separator).

International Oil Pollution Prevention Certificate

Vessels surveyed with acceptable oil filtering equipment will be issued with an International Oil Pollution Prevention Certificate. This survey ensures that the equipment is fitted and operational to the defined limits and confirms that a Shipboard Oil Pollution Emergency Plan (SOPEP) is in place.

MARPOL Annex IV – Regulations for the Prevention of Pollution by Sewage from Ships

Annex IV of MARPOL 73/78 - Regulations for the Prevention of Pollution by Sewage from Ships entered into force on 27th September 2003 and applies to vessels of 200 tonnes and more.

Discharge of Sewage

Discharge of sewage into the sea is prohibited, except:

- Where the discharge of sewage may be necessary for ensuring the safety of the vessel and / or crew or where discharge occurs as the result of damage to the vessel and/or equipment.
- Where the sewage has been comminuted and disinfected using an approved system, at a distance of more than 4 nautical miles from the nearest land, whilst the vessel is en route at a speed of not less than 4 knots.
- Where the sewage has not been comminuted or disinfected, at a distance of more than 12 nautical miles from the nearest land, whilst the vessel is en route at a speed of not less than 4 knots.

Further allowances may be made, for approved equipment, by the relevant Flag State Authority or in accordance with Coastal State requirements.

International Sewage Pollution Prevention Certificate

When in effect, Annex IV will require all vessels to be surveyed and issued with an International Sewage

Pollution Prevention Certificate. This certificate will provide a description of the sewage treatment equipment, communiter, holding tank equipment and discharge connections.

MARPOL Annex V - Regulations for the Prevention of Pollution by Garbage from Ships

Annex V of MARPOL 73/78 - Regulations for the Prevention of Pollution by Garbage from Ships defines garbage as 'all kinds of victual, domestic and operational waste excluding fresh fish and parts thereof'. The provisions of Annex V applies, unless expressly provided, to all ships.

Disposal of Garbage – Within Special Areas

The main distinction with regards to the disposal of garbage, is the area of operation of the vessel.

Special areas have been designated as the Mediterranean, Baltic Sea, Black Sea, Red Sea, the Gulf, North Sea, Antarctic and the Wider Caribbean Region. Within these areas no disposal of garbage is permitted except:

- Disposal of food wastes is allowed as far as practicable from land, but in any case not less than 12 nautical miles from the nearest land.
- However, disposal of food wastes, which have been passed through a comminuter or grinder is allowed as far as practicable from the nearest land, but in any case not less than 3 nautical miles from the nearest land. Such comminuted or ground food wastes shall be capable of passing through a screen with openings no greater than 25mm.

Also, with particular interest to Offshore Support Vessels, the disposal of any materials from fixed or floating platforms and from all other ships when alongside or within the 500 metre zones of such installations.

Disposal of Garbage – Outside Special Areas

Outside the designated special areas and 500 metre zones of installations, the following applies:

- The disposal of all plastics, including synthetic ropes, fishing nets, plastic garbage bags and incinerator ashes from plastic products is prohibited.
- The disposal of dunnage, lining and packaging materials which will float is allowed as far as practicable from the nearest land, but in any case not less than 25 nautical miles from land.
- The disposal of all food wastes and all other garbage including paper products, rags, glass, metal, bottles, crockery and similar refuse is allowed as far as practicable from the nearest land, but in any case not less than 12 nautical miles from land.

Disposal may be permitted if the garbage has been passed through a comminuter or grinder, and as far as practicable from the nearest land, but in any case not less than 3 nautical miles from land.

MARPOL Annex VI – Regulations for the Prevention of Air Pollution from Ships

Annex VI of MARPOL 73/78 - Regulations for the Prevention of Air Pollution from Ships entered into force on 19th May 2005.

The provisions of the regulations, which can be expected to be under review for many years to come, are being enforced due to the harmful effects on the environment of deposits from ship exhausts. Emissions such as sulphur dioxides and nitrogen oxides are known to be a contributory factor in the formation of acid rain and the effects of global warming and the depletion of the ozone layer have led to the legislation.

The regulations define emissions limits for:

- Sulphur Oxides (SOx)
- Nitrogen Oxides (NOx)
- Volatile Organic Compounds

In addition, the regulations prohibit deliberate emissions of ozone depleting substances including halons and chlorofluorocarbons (CFCs).

Further to these limits, Annex VI allows for special

	PLASTIC	**FOOD WASTE**	**GROUND FOOD WASTE**	**DUNNAGE and other materials that float**
In Special Areas	Prohibited	Permitted as far as possible from land and never closer than 12 miles from shore	Prohibited	Prohibited
Outside Special Areas	Prohibited	Permitted if more than 12 miles from shore	Permitted if more than 3 miles from the nearest land	Permitted if more than 25 miles from shore
Within 500m of fixed or floating installations	Prohibited	Prohibited	Prohibited	Prohibited

Figure 14.1   Summary of Garbage Disposal

Sulphur Oxide (SOx) Emission Control Areas (SECAS) to be established. Within these special areas the sulphur content of fuel used on onboard vessels must not exceed 1.5% m/m. Alternatively, an exhaust cleaning system may be used to maintain the 1.5% m/m limit within these areas. It should be noted that the North Sea has been designated as a SECAS since July 2005.

Shipboard Incinerators

The regulations also specify restrictions for the use of shipboard incinerators. Incinerators installed onboard after 1st January 2000 must be type approved and any use of non type approved incinerators installed after this date is prohibited.

Incinerators installed onboard before this date need not be type approved according to IMO Resolution MEPC 76(40). However, incineration of PVC should not take place in incinerators that are not type approved.

Incineration of certain substances is prohibited such as:
- Polychlorinated biphenyls (PCBs)
- Garbage containing traces of heavy metals
- Refined petroleum products containing halogen compounds
- Residues from MARPOL Annex I, II and III cargoes

The use of an incinerator during port calls, in an estuary and when at anchor close to a port are prohibited by the regulations.

International Air Pollution Prevention Certificate

On completion of the initial survey, all vessels of 400 gross tonnes and above will be issued with an International Air Pollution Prevention Certificate.

Record Keeping Requirements (MARPOL)

- Every vessel of 12 metres or more in length shall display placards informing the marine crew and project personnel of garbage disposal requirements in compliance with MAPROL.
- Every vessel of 400 gross tonnes or above, and every vessel certified to carry 15 passengers or more, shall carry a garbage management plan and maintain a garbage record book. The garbage record book should be kept for a period of at least two years after the last entry is recorded. The garbage management plan should include procedures for the collecting, storing, processing and disposal of the garbage.
- The garbage record book must be maintained up to date at all times and all details must be recorded on each disposal of solid waste, whether into the sea, to a shore reception facility, waste carrier or incinerated onboard. These details should include the date and time of the discharge or incineration, position of the vessel or name of the port facility / waste carrier, category of waste discharged or amount incinerated (m3) and must be signed off by the officer in charge of the operation.
- Every vessel of 400 gross tonnes or above is required to maintain an Oil Record Book Part I (Machinery Spaces) and should be kept for a period of at least three years after the last entry was made.
- The Oil Record Book Part I (Machinery Spaces) must be maintained up to date at all times and details recorded whenever operations such as the discharge of sludge, ballasting or cleaning of fuel tanks, discharge of dirty ballast from fuel tanks and the discharge or disposal of bilge water from machinery spaces take place. The time of discharge or transfer, position or port at the time of the discharge or disposal and the quantity disposed of shall be recorded in the Oil Record Book and shall be signed off by the officer in charge of the operation.

Garbage Management (Environmental Protection Act 1990)

MARPOL details the requirements for disposal of garbage at sea, however it does not provide details for the proper disposal of residues and effluent to shore reception facilities.

The Environmental Protection Act 1990, imposes a 'duty of care' on all persons concerned with the transfer of controlled waste. This 'duty of care' applies to any person who produces, imports, carries, keeps, treats or disposes of controlled waste, or as a broker has control of such waste. The requirements of the regulations therefore apply to the Offshore Support Vessel.

Clear statements regarding the compliance of all parties are detailed in the guidance. Duties and requirements for the vessel crew have been summarised below:

Vessel Master

It is the Master's responsibility to ensure that all 'Waste Producers' onboard their vessel conduct their duties as required by the Environmental Protection Act 1990 and MARPOL Annex V 73/78.

Waste Producers

The onboard waste producers are responsible, together with the Master, for the care of their waste while it is onboard their vessel.

Waste Producers are responsible for ensuring that waste onboard the vessel is correctly identified and segregated and that the description of the waste which is exported from the vessel is accurate.

They must ensure that the information provided for the Shore Carrier and Shore Base includes all data necessary for the safe handling, disposal, treatment and recovery of the waste.

Waste Management Controller

In order to ensure that waste is correctly segregated and designated prior to discharge from the vessel, the Master or Vessel Owner should appoint a designated Waste Management Controller. The Waste Management Controller should report directly to the Master.

The Waste Management Controller must ensure that all waste produced onboard the vessel is suitably segregated, packaged, labelled and described prior to transfer to the shore based carrier. It is the Controllers responsibility to ensure that the carrier completes all form sections required.

It is the duty of the Waste Management Controller to ensure that the Waste is correctly identified and the appropriate 'Consignment Note' (Controlled or Hazardous) is correctly completed with all known details provided.

The Waste Management Controller is responsible for ensuring that all Consignment Notes are distributed correctly and that a record of all Consignment Notes is maintained.

Classification of Waste

Waste transferred ashore can be defined as either 'Controlled Waste' or 'Hazardous Waste'. Differing forms are utilised when transferring either classification of waste and it is therefore essential that an onboard assessment is made to classify the waste.

Classification of the waste onboard the vessel can be assessed in a three stage process. The conclusion of the three stage process will result in the classification of the waste as either 'Controlled Waste' or 'Hazardous Waste'.

An example assessment process is summarised below:

Stage 1

The initial stage of the Waste Assessment Process is to determine the broad classification of the waste and to ascertain whether the waste is defined as 'Hazardous Waste'.

Stage 2

The second stage of the Waste Assessment Process is a closer examination of the waste with regard to possible or actual hazardous properties.

Stage 3

The third stage of the Waste Assessment Process is an examination of any waste which cannot be identified or classified in the specified data sources. Merchant Shipping Notice M1678 should be consulted for a full list of substances, however the following tables provide some basic guidance on what is classed as 'Controlled Waste' and 'Hazardous Waste'.

Controlled Waste	Hazardous Waste
Wood	Fibre-Optic Cable
Steel	Oily Rags
Plastics	Waste Oil (Diesel / Gas Oil)
Empty Paint Drums	Lubricating Oil
Shot Blast	Cooking Oil
Paint Chippings (non-lead)	Grease
Rope	Aerosol
Wire	Nicad Batteries
Electrical Cable	Lead Acid Batteries
Paper	Paint Chippings (Lead based paint)
Cardboard	Full Paint Drums
Concrete	Chemicals (Degreasers etc)
Food Waste	Hydraulic Hoses

Figure 14.2 Controlled and Hazardous Waste

Once the waste has been classified, the waste management controller can determine whether a 'Controlled Waster Transfer Note' or a 'Hazardous Waste Transfer Note' should be utilised.

Transfer of the Waste

Waste shall only be transferred to an authorised person or persons from a registered Waste Carrier and the final disposal site must be in possession of a Waste Management License relevant to the waste being transferred.

Anyone intending to transfer waste must ensure that the Waste Carrier is registered and that the final disposal site is licensed. This is the responsibility of the waste producer (vessel).

A registered carrier's authority for transporting waste is either his certificate of registration or a copy of his certificate of registration.

The certificate or copy of the certificate will show the date on which the carrier's registration expires.

All copy certificates must be numbered and marked to show that they are copies.

In all cases other than those involving repeated transfer of waste, the holder should ask to see, and should check the details of, the waste carrier's certificate or copy certificate of registration.

Record Keeping (Environmental Protection Act 1990)

Controlled Waste Transfer Notes should be maintained onboard for a minimum of two years and Hazardous Waste Transfer Notes for a minimum of three years.

International Standards Organisation (ISO) 14001

The requirements of ISO 14001 are not mandatory for vessels or for shipping companies, however there is an expectation from many oil companies that operators and ship owners should comply.

The requirements for compliance with ISO 14001 are extensive, however the main areas of focus can be categorised as documentation, implementation and maintenance and improvement of the environmental management system, required by the legislation.

Documentation of the Environmental Management System

- Establish an Environmental Policy.
- Identify company operations that have a significant environmental aspect and provide suitable documented procedures.
- Identify and document a management process for potential situations and accidents that could have a significant impact on the environment.
- Identify applicable legislation and provide a documented system of compliance with such legislation.

Implementation of the Environmental Management System

- Provide a suitable organisation including personnel resources and applicable training to implement and then maintain the environmental management system.
- Identify objectives and targets to reduce environmental impacts due to the activities and operations of the company.
- Ensure awareness of the Environmental Management System within the company.

Maintenance and Improvement of the Environmental Management System

- Maintain records of environmental emissions and assess opportunities for reduction of emission levels through continuous improvement.
- Evaluate compliance with mandatory regulations and the company environmental management system.
- Perform regular internal management audits and management reviews in order to highlight deficiencies in the Environmental Management System and assess areas for improvement of the system.

Chapter 15

BALLAST MANAGEMENT

General Introduction

The International Maritime Organisation Assembly Resolution A.868(20) — Guidelines for the control and management of ships ballast water to minimise the transfer of harmful aquatic organisms and pathogens, details recommendations for the implementation and operation of an onboard Ballast Management Plan.

The function of this Ballast Management Plan is to assist the vessel crew in complying with the quarantine measures which are intended to minimise the risk of transplanting such organisms from ship's ballast water into differing environments, whilst maintaining the integrity of the vessel.

There are thousands of marine species that may be carried in ships' ballast water; basically anything that is small enough to pass through a ships' ballast water intake.

Figure 15.1   Harmful Aquatic Organisms and Pathogens

Although the Resolution has not been adopted by all nations, certain provisions have been adopted by various national and coastal states. Therefore, Ballast Management Plans are required to be onboard and working effectively for any vessel visiting, for example, the United States or Australia.

These national and in some cases, coastal requirements, such as in the Orkney Islands, have established controls on the discharge of ships' ballast water that will minimise the potential for colonisation of their rivers and estuaries by non-native species. The preferred option is mid-ocean ballast water exchange prior to arrival. Accordingly, the nations and coastal states most concerned have provided guidance to ships for ballast management. However, it should be noted that the full IMO Regulations are in force from January 2009.

Requirements and Reporting

The main concern for the national and coastal states, regarding ballast water management, will be to be assured that not only is a suitable Ballast Management Plan in place, but more importantly that the vessel crew are aware of the recommendations and that the requirements of the plan are being followed and adhered to.

Concerned countries have therefore introduced a requirement which, though not standard, generally requires vessels to report in advance, with details of the quantity of ballast water onboard, the place of origin of the ballast water and confirmation that a ballast management procedure has been followed. In most cases it is mandatory to make the report, even though the actual ballast exchange in midocean (or other management procedure) remains voluntary.

As such, the vessel will require to maintain a full and accurate ballast log. Even if a ship is not trading in an area where ballast water information is required, it may later prove worthwhile to have a history of what water has been carried.

These records are the main proof of compliance with the requirements and therefore the maintenance of such records cannot be stressed enough when operating in waters where these regulations have been adopted and are enforced.

Safety Considerations

It should be noted that, although the exchange of ballast on-route to a port or area, is a requirement to protect the environment, the safety of the vessel and thus the crew onboard must always remain the main consideration. The main requirements of any Ballast Management Plan must therefore provide a sequence of ballast exchange that at no time compromises the safety or integrity of the vessel.

The safety points outlined below are intended to emphasise that the consequences of an inadvertent error at sea can be more significant than the same error made in port.

Assessment of Conditions prior to commencement of ballast water exchange operations

It is accepted that certain conditions and circumstances will arise whereby a planned ballast water exchange, at sea, would be considered unsafe. It is therefore essential that no ballast water exchange operations should commence until a full review has been made of the vessels current status including any expected or possible changes to this status during the planned duration of the operations.

The following factors should, at a minimum, be considered as critical with regard to any possible ballast water exchange operations.

- Weather conditions: both prevailing and forecast. Weather conditions should include, but not be limited to, sea state and swell, wind force and direction, ice conditions and area (i.e. Hurricane or TRS zones). A period of suitable weather conditions should be identified for the full ballast water exchange operations to be completed without interruption.
- Vessel Type: For Offshore Support Vessels, vessel type considerations should not pose much concern with regards to performing ballast water exchanges at sea. However, all possible areas of concern should be assessed including trim, stability, cargo onboard and safety of personnel.
- Vessel Location: When loading ballast, every effort should be made to ensure that only clean ballast water is loaded and that the uptake of any sediment is minimised. This will obviously reduce the possibility that harmful aquatic organisms and pathogens. Similarly, areas where there is a known outbreak of diseases, communicable through ballast water, or in which phytoplankton blooms are occurring, should be avoided wherever practicable as a ballast source.
- Crew Availability: Ballast water exchange operations should not be commenced until sufficient personnel are available to conduct the operation until completion. Personnel involved in the operation must have the appropriate knowledge, competency and experience to perform their specified function.
- Vessel Operations: No ballast water exchange operations should be conducted in tandem with any other shipboard or project related activities that may interfere with the safe conclusion of the ballast water exchange operation.
- Personal Safety: No ballast water exchange operations should be conducted in circumstances whereby the safety of the vessel, personnel or environment would be in any way compromised.

It must be stressed that the vessels own Trim and Stability Booklet must be referred to at all times, and its overriding importance stressed. Equally, shear forces, stress limits and bending moments must be taken note of. The Trim and Stability Booklet should be consulted to determine these limits, and be regarded as the definitive document.

Safety Considerations during ballast water exchange operations

In conducting ballast water exchange operations, the following operational safety considerations should be taken into account:

- Avoidance of over and under pressurisation of ballast tanks.
- Free Surface effects on stability.
- Current and predicted weather conditions.
- Weather routeing in areas seasonally affected by cyclones, typhoons, hurricanes or heavy icing conditions.
- Maintenance of adequate intact stability in accordance with an approved trim and stability booklet.
- Permissible sea going strength limits of shear forces and bending moments in accordance with an approved loading manual.
- Torsional forces, where relevant.
- Minimum / Maximum forward and aft draughts.
- Wave induced hull vibration.
- Documented records of ballasting and/or de-ballasting.
- Contingency procedures for situations which may affect the ballast water exchange at sea, including deteriorating and unforeseen weather conditions, pump failure and loss of power.
- Time to complete the ballast water exchange or an appropriate sequence thereof, taking into account that the ballast water may represent a large percentage of the total cargo capacity for an Offshore Support Vessel.
- Any other systems that are interconnected with the ballast system should be isolated, such as bilge water arrangements. This will ensure that no inadvertent pollution is possible as a result of ballast water exchanges.
- It is as important to avoid under pressure in a tank due to emptying, as it is to avoid overpressure when filling. The consequences of bulkhead damage, or even tank collapse, at sea will be even more significant than in port.

Crew Training and Familiarisation.

In order to ensure that any ballast water exchange operations are conducted efficiently and safely, an important area of consideration is training and familiarisation. This aspect of any ballast water exchange operation should include, as a minimum, the Master, Chief Officer, 2nd Officer and Engineer(s) involved and should cover the following:

- The vessel's pumping plan including ballast pumping arrangements.
- Positions of associated air and sounding pipes.
- Positions of all appropriate compartment and tank suctions and pipelines connecting them to ship's ballast pumps.
- In the case of use of the flow through method of ballast water exchange, the openings used for release of water from the top of the tank together with overboard discharge arrangements.
- The method of ensuring that sounding pipes are clear, and that air pipes and their non-return devices are in good order.
- The expected times required to undertake the various ballast water exchange operations.
- The methods in use for ballast water exchange at sea with particular reference to required safety precautions.
- The method of onboard ballast water record keeping, reporting and recording of routine soundings.

Methods of Managing Ballast Water

There are two recognised methods of conducting out ballast water exchange at sea:
- The sequential method in which ballast tanks are pumped out and then refilled with water.
- The flow-through method in which ballast tanks are overfilled by pumping in additional water to dilute the original water.

Alternatively, sediment removal (or reduction) or retention of ballast water onboard a vessel may be an option that can be considered.

Sequential Method

The sequential method is a straight forward empty-then-refill procedure. The process requires the removal of large weights from the vessel (i.e. full ballast tanks) in a dynamic situation, and then their replacement. A relatively simple procedure, for a vessel in port, this method has obvious dangers whilst the vessel is at sea.

To utilise this method, a step by step plan should be put in place with the action to be taken (fill or empty) and the tanks involved for each step of the methodology provided. Guidance should also be provided, detailing the assumed weight of fuel and domestic drinking water onboard, estimated draughts, bending moments and shear forces at each stage of the empty-refill methodology. With such a methodology, it will be noted that the original condition is restored after each pair of steps (empty and refill). This step by step approach therefore allows a positive decision to be made at that time, taking account of the ship's position, weather forecast, machinery performance and degree of crew fatigue, before proceeding to the next pair of steps. If any factors are considered unfavourable the ballast exchange should be suspended or halted.

Flow-through Method

The flow-through method, whereby tanks are overfilled by pumping in additional water, has the advantage that it can be used in weather conditions which would be marginal for use of the sequential method, since there is little change to the condition of the ship. However, the flow-through method introduces certain other risks and problems, such as over pressurisation of the tanks, which must be considered before using this procedure.

Sediment Removal or Reduction

Where practical, cleaning of the ballast tanks to remove sediments can be considered a viable option. The requirement to perform sediment removal or reduction will be dependent on known sediment levels, vessel location and local conditions. Flushing by using water movement within a tank to bring sediment into suspension will only remove a part of the mud, depending on the configuration of an individual tank and its piping arrangement. Removal may therefore be more appropriate on a more routine basis during scheduled dry-dockings.

Any removal operations requiring tank cleaning, should be conducted in accordance with the vessel's Permit to Work system and should only be carried out after a Risk Assessment has been completed.

Ballast Retention

The retention of ballast water onboard the vessel may be considered a more suitable option to continual ballasting and de-ballasting during cargo operations, to maintain trim and list.

If possible, internal ballasting should be conducted in preference to discharging of ballast water. As with ballasting and de-ballasting operations, records should be maintained onboard regarding any internal ballasting operations.

Appendix A

A BRIEF HISTORY OF SATURATION DIVING SYSTEMS AND ROVS

Introduction

The saturation diving systems that are fitted onboard modern Dive Support Vessels have been developed over a considerable period of time. The very first rudimentary diving bells are believed to have been developed over 2,000 years ago with the earliest recorded system being attributed to Edmund Halley in 1691.

However, the main period of development from these basic systems to the sophisticated systems now in operation, can be considered to have commenced in 1897 with the launch of the submarine *USS Holland*.

Although not the first to have been built, the *USS Holland* was the first successful submarine to be constructed and operated. Bought by the US Navy in 1900, from the original private owner, John P. Holland, the vessel spent extensive periods on trials which provided the US Navy with valuable experience and knowledge. The comparative success of the *USS Holland* led to the development of US Navy funded submarines, but not without problems.

The loss of the *F4* on 25th March 1915 was the first major disaster for the US Navy. The submarine exploded and sank in some three hundred feet of water off Honolulu harbour with the loss of all twenty one submariners. Following several salvage attempts, the vessel was successfully re-floated, allowing the US Navy to examine the vessels failings which were to be essential as submarine operations increased and further accidents occurred. Such an accident befell the *Squalus* on 23rd May 1939. However, the subsequent rescue of thirty three of the crew was seen as a major breakthrough.

Lost in some two hundred and forty three feet of water, the survivors of the initial sinking would have known that no-one had ever survived such a disaster. However, within hours of the incident, the sister vessel *Sculpin* was on location, followed by a number of rescue vessels. A new development, The McCann Rescue Chamber was utilised to recover the stricken sailors to the surface.

In 1953, the first free-diving submersible *Trieste I* was launched and on 23rd January 1960, dived to a world record depth of 35,800 feet in the Marianas Trench. The *Trieste I* was a bathyscaphe capable of descending and ascending independent of any third party vessel. Ten years after this dive, the submersible was used to locate the sunken nuclear submarine *USS Thresher*.

Although the losses of the *F4* and the *Squalus* were major blows to the US Navy and its ongoing submarine development program, the loss of the *USS Thresher* on 9th April 1963 was a major catalyst to progress of saturation and habitat diving.

All one hundred and twenty nine crew were lost as a result of an initial implosion onboard the submarine and the subsequent sinking of the vessel in some eight thousand four hundred feet of water. No means were available, at the time, to attempt a rescue in such water depths.

The *Trieste I* had recently been moth balled, but as a direct consequence of the loss of the *USS Thresher*, the vessel was returned to service and deployed to the site to ascertain the reasons for the loss. The information gained from numerous dives to the wreck location and the subsequent investigations into the loss of the *USS Thresher* convinced the US Navy to form the Deep Submergence Systems Project. The purpose of the Project was to develop deep water capabilities.

One of the first decisions made by the committee was to design and construct a second bathyscaphe, *Trieste II*. This submersible would, in 1965, be used in a similar capacity as its predecessor to investigate the loss of the USS Scorpion.

In addition to the ongoing submersible program, a major development was the focus on underwater habitats which would eventually cross over into the offshore oil industry.

Experimentation in hyperbaric environment living began in the early 1930s. The concept of saturation diving and living in underwater habitats was introduced by G. Bond, a U.S. Navy submarine medical officer. However, it was not until the early 1960s that Jacques Cousteau worked privately on an underwater habitat and in 1963, five divers spent a period of one month in the *Conshelf II* habitat at a water depth of thirty six feet. *Conshelf III* followed, with a team of six divers living in the unit for a period of six days at a depth of three hundred and twenty eight feet.

The *Sealab I* was a ten metre long living chamber, operated by the US Navy, which was trialled off the coast of Bermuda in 1964. The unit was stationary on the seabed with communications, breathing gas and emergency supplies provided by a combination of a support vessel on the surface and shore connected cables and umbilicals. The trials were a complete success with a four man team of divers remaining on the seabed for a period of ten days at a depth of some one hundred and ninety two feet. During this period, the divers were able to exit the living habitat for test dives and returned successfully to the living chamber.

In 1965, the *Sealab II* expanded the initial trials with a test period at a depth of two hundred and six feet. Again the duration of the trial was set at ten days, however by utilising a team rotation system, the habitation of the living chamber continued for some weeks. Ex-astronaut Scott Carpenter stayed in the habitat for thirty consecutive days. Further depth increases were made with *Sealab III* in 1969.

The Sealab and Conshelf habitats were successful in that they proved that human life could be supported for extended periods of time on the seabed in controlled living chambers. However, such studies relied on the personnel remaining in the same place and at the same depth for the entire period. For practical applications, the need to mobilise the personnel to particular locations for a specific operation and then to transfer them to another location, possibly at a different depth, was of major importance.

The importance of this requirement was to lead to significant progress in the 1960s, 1970s and 1980s as the offshore oil and gas industry proved that there were commercial possibilities in the use of saturation diving. The oil and gas industry therefore provided a shift from theoretical and experimental trials to profitable operations.

The earliest systems in the oil and gas industry were modular units fitted on a temporary basis on offshore platforms or barges. The use of such systems had the same limitations as the earlier habitat systems in that they were often restricted to one location and therefore the obvious next stage was to fit saturation diving systems to monohull and semi-submersible vessels. Initially, dive systems where fitted to converted supply vessels and it was not until the 1980s that purpose built dive support vessels started to enter the market.

A Brief History of the ROV

The use of manned submersibles and saturation diving bells for underwater operations has many disadvantages due to the constraints imposed by using human beings to control and operate the systems. As such, the use of remotely operated vehicles (ROV) can be seen as a more suitable option where human intervention is not essential. The development of the ROV can therefore be considered to have stemmed from advances made in submarine design.

There are two people generally credited with the development of the first remotely operated vehicles. In the mid 1860's Captain Giovanni Luppis of the Austrian Navy designed a robot boat that could carry explosives to an enemy vessel. The boat was designed to be powered by steam or clockwork and steered by cords trailed behind it and ultimately led to the development of the first torpedo. However, it was not until 1953 that the first ROV in the modern sense was developed. It was named *Poodle*, and evolved from an underwater scooter created by an underwater photographer named Dimitri Rebikoff.

It was the US Navy that advanced these concepts by developing robots to recover ordnance lost during tests at sea.

The next step in advancing the technology was performed by commercial firms that saw the future in ROV support of offshore oil and gas operations. The transition from military use to the commercial world progressed quickly and has now reached a level of cost effectiveness that allows organisations from police departments to academic institutions to utilise them. They are now even being marketed as premium items for recreational boat owners.

INDEX

A

abandonment and recovery 63
    arrangement 56, 60, 62
    operations 92
    winches 94
    winch wire 91
    wire 91
accident 110, 130
    and incident
        investigation 117, 131
        reporting 110, 117, 130
accommodation 11
    ladder 133
acoustic
    and tracking sensor 41
    systems 6
active
    heating system 28
    heave compensation 81, 82
actual breaking load 86
adaptation of procedures 119
adverse weather
    conditions 37
    policy 110
A-Frame 12, 46, 100
    type 37
aligners 58
    and straighteners 56, 57
    track 58
anchor 69
    pattern 68
annual DP trials 102
anode replacement 66
anti-heeling system 47, 48
approach rollers 90
armoured umbilical 45
arrival at a safe haven 71
as-built survey 93
as-laid survey 78
audit 110
    internal 107
auditee 107
auditor 107
auto
    depth 38, 39
        control 41
    heading 38, 39, 41
autonomous underwater vehicle 38, 43
avoidance of risk 119
azimuth thruster 7, 8, 100

B

back-tension 61
ballast 5
    and bilge system 10
    management 151
        plan 151
    retention 153
    water 82
    water exchange 151
        safety considerations 152
ballasting, stability and watertight
    integrity 110
banksman 84
bellman 25
bending
    moment 51, 92
    radius 55, 90
boom
    angle 82
    box type 51
    lattice type 51
    length 51
    tip 51
bow tunnel thruster 8
breathing gas 20, 26
    management 22
bridge design 8, 9, 10
buckling (pipe) 55
built-in breathing system
    (BIBS) 19, 20, 22, 34
    face mask 22, 26
bunkering 110
buoyancy
    and ballast control 40
    chamber 41
    module 40

C

cable 37
    burial 38, 41
cameras, video functions and
    lighting 40, 41
CAP 437 13
cargo
    operations 110
carousel 6, 53, 55, 90, 93
    spooler 61
catenary 59, 69
Centurion HD (workclass ROV) 39
certificate of medical fitness 65
chamber 1
    depth 23
    pressure 23
    temperature 23
clamping mechanism 19
classification of waste 148
Classification Society ... 13, 17, 47, 100, 101
cleaning
    and debris removal 74
    operations 66
clump weight 93, 98
CO_2
    absorber 20, 23
    scrubber 28
Code of Practice
    for the Safe and Efficient Operation of
        Remotely Operated Vehicles 73
    for the Safe Use of Electricity
        Underwater (AODC 035) 78
Code of Safe Working Practices for
    Merchant Seamen 133, 136, 140
colour video camera 42
communication system 20, 23, 31, 32
Company Security Officer 113, 114, 115
compression chamber 20
constant tension 81
    mode 81
construction
    operations 47, 81, 95
    support 1, 66
    vessel 47
        crane classification 47
control
    line 54, 55
    of discharge of oil 145
    system 95, 96
controllable
    and fixed pitch blades 8
    pitch propeller 7
controlled waste 148
crane 5, 10, 47, 100
    boom (or jib) 49
    boom type
        with lattice arrangement 47
    box boom type
        with luffing cylinders 51
    capacity curves
        in graphical format 50
        in tabular format 49
    controls 5
    failure modes effect and analysis
        (FMEA) 83
    hook 70
        latch 70
    knuckle boom 49, 51
        in stowed position 52
    lattice boom type 50, 51
        with luffing wires 48
    modes and functions 81
    operational considerations 47
    operations 2, 10, 70, 98, 110
    operator 76
    overload alarm 82
    position 47
    telescopic 49, 52
    types 49
    vessel 47
    wire 2, 5, 70, 81, 84
        certificate 85
        installation 86
        selection 85
currents and tidal conditions 75
    surface and sub surface 74
cursor
    clamp arrangement 29
    latch 44
    system 44
    wire 44

D

damage
    and condition surveys 66
    stability 53
data collection 38
dead man anchor 71, 93

debris
   inspection 66
   removal 37, 78
deck
   carousel 90
   mounted reel 56, 60, 61
      tie-in arrangement 57
   radius controller 60, 62
   reel 90
decompression 67
   sickness 21, 67
   tables 67
Defined Levels of Authority and Lines
   of Communication 109
departure angle 57
depth monitoring 31
design approval 47
Designated Person Ashore (DPA) .. 107, 109
Det Norske Veritas (DNV) 100
Differential Global Positioning
   System (DGPS) 6, 96, 97
discharge 78
   of sewage 145
disposal of
   food waste 146
   garbage
      outside special areas 146
      within special areas 146
        diver 2, 6, 9, 37, 77, 81
   emergency 15
   heating 28
   excursion umbilical 26, 27, 69
   gas supply 26, 27
   intervention 37, 66, 69, 70
   intervention operations 6
   observation 37, 74
   personal equipment 18, 35
   umbilical 26, 74
diverless latch 93
diving 8, 10
   at Work Regulations 1997 65
   chamber 2
   contractor 65
   helmet 36
   in the vicinity of pipelines and
      wellheads 65, 70
   Medical Advisory Committee
      DMAC 26 17
   operations 1, 2, 65, 95, 98, 110
      associated with lifting operations. 65, 70
      from vessels operating in dynamically
        positioned mode 65, 67
      within anchor patterns 65, 68
   personnel 11
   planning 65
   project plan 65
   supervisor 65
   support 1
   system 2
   vessel 10
diving bell 1, 2, 15, 16, 17, 18, 24,
   .... 48, 68, 77
   atmosphere 70
   ballast release system 27
   construction 25
   contamination 70

control 9, 17, 68, 76
   room 18, 31, 32
gas supply 26
handling system 1, 2, 15, 18, 29
hangar 29
hoist wire 25
launching system 2
life support system 18
life support systems and functions 24
main
   lift wire 29
   umbilical 26
mating 20
   clamp 23, 29
planning 66
positioning 24
project plan 67
supervisor 68, 77
support 95
vessel 1, 2, 5, 9, 15
   assistance 72
system
   chamber complex 18
   class notations 16
   trunking 22

DNV classification rules 47
Document of Compliance
   (DOC) 107, 108, 110
DP 6
   Class
      DNV 101
      IMO 101
      Lloyds 101
      NMD 101
   classification 100
      DNV and NMD 101
      (Lloyds Register) 101
   control system 100
   familiarisation 103
   mode 9, 10, 90
   operations 102
   operator 96
   operators 103
      certificate 103
   reference system 6
   reference unit 98
   system 93, 95, 96, 100, 102
dredging operations 66
drilling support 37, 74
dropped objects 67, 70, 74, 89
drop test 33
DSV
   Class I 16
   Class II 16
   Class III 16
duties
   of employees 118
   of the employer 117
   of the management company 118
dynamic
   loads 43
   positioning 5, 10, 95
      operations 10
      purposes 96
      system 1, 95, 96

Vessel Owners Association
   (DPVOA) 103
DYNPOS
   AURT 101
   AUT 101
   AUTRO 101
   AUTS 101
   T 101

E

EERVs 35
emergency
   and contingency situations 73
   evacuation 15
      system 1, 18, 32
   location equipment 76
   response procedure 115
   situations 10, 65
   situations procedure 110
   station 94
   system 31, 32
enclosed space
   definition 138
   hazards 138
end termination operations 63
engine room
   procedures 110
entry into enclosed spaces 117, 138
environmental
   analyser 20, 23
   control 23
      system 20
   forces 95
   limits
      for monohull vessels 16
      for semi-submersible vessels 16
   management 145
      system 149
   policy 149
   reference system 96, 100
   reference systems 95
   sensor 95
equipment lock 19
ergonomic design 120
escape trunk 35
evacuation trunking 71
evaluation of unavoidable risk 119
example pipe lay systems 55
excursion umbilical 43, 68, 74, 77
exit
   arrangement 53
   monitoring frame 56, 57, 59
   roller 91
   roller box 56, 57, 59
external audit 110
   Classification Society 110

F

failure modes and effects analysis 54, 101
false cores 60
familiarisation 110, 132
fanbeam 96, 97
   operation 100
Finding (FN) 107
fire deluge system 23

fire-fighting system 20
first aid assistance 65
fixed
    ballast....................................41
    payload 41
flag State
    administration....................108
    Authority 47, 53, 115
    legislation...........................108
    verification and acceptance document 102
flange 78
fleet angle 86
fleeting 57, 90, 91
flexible..55
    pipelines and flowlines 54
    product 93
floating production storage and
    offloading (FPSO) vessel 69
flooded capsizing test....................33
flow-through method 153
FMEA (DP) 102
food lock 34
forces involved in dynamic positioning....95
free swimming option 39
functional requirements
    of the ISM Code 108
    of the ISPS Code................113

G

gangway 133
    and accommodation ladder access 133
    use, security and piracy........110
garbage
    management
        (Environmental Protection Act
        1990) 145, 147
    plan......................................147
    record book.........................147
gas
    distribution 22
    reclaim bag (tank) 22
    regeneration 20, 22
    regulation 31
    storage...................................22
    and distribution 20, 21
gauntlet manipulator....................39
generic DP system 95
gimballed sensor head.................98
global positioning system (GPS).............96
GPS
    fix 97
    receiver................................96
grit cleaning 67
gross overload protecting system........81, 82
guide
    weight 29
    wire
        system 29
        weight and winch arrangement 29
Guidelines
    for the Design and Operation of
    Dynamically Positioned Vessels 54, 68
    for the Issue of a Flag State Verification
    and Acceptance Document................102
    for Vessels with Dynamic Positioning

Systems 102
guide pads 61
gyro..100

H

hang off clamp 62, 63
hazardous waste.........................148
heading reference system 95, 96, 100
Health and Safety
    at Work Act 1974..................65
    at Work Regulations...........137
    Executive (HSE).................65
    Executive (HSE) Lifting Operations and
        Lifting Equipment Regulations
        1998 141
heating and emergency heating system 27
heave 95
    compensated cranes..............81
    compensation 5, 81
heavy lift 82
    operations 5
helicopter 13
    Certification Agency (HCA) 13
    landing officer 14
    operations 100, 110
    pilot 14
helideck................................12, 15
    D-value.................................13
    structural strength 13
heliox 21
helium 21
Herald of Free Enterprise...........107
Hercules (workclass ROV).........38
HiPAP (High Precision Acoustic
Positioning)...................................97
horizontal
    clearance 69
    (pan) adjustment..................41
hot water diving suits 35
HPR (Hydroacoustic Positioning
Reference) 97
HSE Diving Information Sheet
    No. 3 22
    No. 4 20
human
    element 121
    factors and the human element ..117, 120
    factor statistics....................120
    intervention 73, 77
humidity 23
    control 31
hydro-acoustic position reference
    system 96, 97
hydrocarbons..............................70
hyperbaric
    evacuation 65, 71
    evacuation system................17
    lifeboat.........................2, 18, 33, 71, 72
    chamber 34, 71
    control panel.......................31
    embarkation to and launching of 71
    launch control position 35
    lifting beam.........................34
    marking (D027)..................33
    mating 33

transit to a safe haven 71
rescue centre.....................34
toilet 35
welding 66
hyperoxia...................................22
hyperoxic...................................20
hypoxic......................................20

I

IMCA 33, 104, 105
    M117 Training and Experience of
    Key DP Personnel 103
    M166 Guidance on Failure Modes
    and Effects Analyses 102
    M171 Crane Specification Document..86
    organisation 104
IMO 33, 101, 102, 114, 115, 121
    guidelines...........................100
    Resolution A.534 (13) Code of Safety
    for Special Purposes Ships 53
impact test..................................33
incident........................110, 130
    (hazardous occurrence).......107
    investigation......................129
    reporting............................110
induction 117, 132
initial and periodic examination,
    testing and certification of ROV
    handling systems 46
inspection...................................66
installation
    inspection.......................37, 74
    of mats 66
integrity check.......................66, 76
interface tools............................42
intering
    ballast system 82
    system 81, 82
internal audit............................110
internal safety systems 10
International
    Air Pollution Prevention Certificate....147
    Association of Offshore Diving
    Contractors (AODC)................103
    Code of Practice for Offshore Diving ...66
    Labour Organization (ILO)..................47
    Marine Contractors Association
    (IMCA)......................................73, 103
    Maritime Organization 107
    Assembly Resolution A.868(20)
        Guidelines for the control and
        management of ships ballast water
        to minimise the transfer of harmful
        aquatic organisms and pathogens ...151
    Oil Pollution Prevention Certificate ...145
    Sewage Pollution Prevention
    Certificate 145
    Ship and Port Facility Security
    (ISPS) Code 113
    Standards Organization (ISO) 14001.149
ISM
    Code 107, 108, 110
    certification......................................108
    internal audit 111
ISPS 113

Code 9, 114, 115

J

jetting..67
jib..50
J-lay..54
   installation methodology.....................55
joint 78
jointing 63

K

kicker tread 135
knuckle boom crane 86

L

lateral bending...................................57
lattice
   boom 50
   jib arrangement 50
launching and recovery control station 32
lay down...53
   head 94
   position 84
   process 94
lay system 89
L-drive 7
lead
   auditor 107
   string 91
lifeboat chamber................................35
life support..1
   package 71, 72
lift
   bags 71
   path 84
   planning....................................81, 83
   wire winch 45
lifting
   appliances......................................47
   beam 72
   capacity and working radius 48
   operations 70, 84, 86
      and lifting equipment......................117
lightweight framing............................39
limits and cut offs........................81, 83
limit switch 87
living chamber.................15, 16, 17, 19, 23
   complex 18
      life support systems and
         functions 18, 20
loading
   and spooling operations 90
   path 70
   tower 60, 61, 62
      tensioner track 62
location and recovery operation...............74
locked gate type hooks 70
lower and upper chute arrangement 62
low light camera 42
luffing
   cylinders...50
   wires 50
luff (lift and lower) 50

M

main
   lift wire 44
   propeller
      manoeuvring system......................8
      propulsion.......................................7
   propulsion...........................2, 3, 6, 8
      system 7
   umbilical system............................42
main bell
   lift wire 31
   and winch arrangement.................29
   winch system.................................29
maintenance.....................................110
Major Non-Conformity (MNC) 107
management
   review....................................110, 111
manifolds 37
manipulation
   of subsea objects 42
   of valves 66
manipulator 37, 38, 42, 76
   arms 38, 39
   interface tools and skids 40, 42
   operations 39
manoeuvring 9, 83
   and propulsion system....................2
   system 6, 95, 96, 100
      propeller and rudder 7
manual handling 117, 141
marine growth 78
   surveys 66
marking of gas cylinders 21
MARPOL
   Annex I
      Regulations for the Prevention of
         Pollution by Oil 145
   Annex IV
      Regulations for the Prevention of
         Pollution by Sewage from Ships 145
   Annex V
      Regulations for the Prevention of
         Pollution by Garbage from Ships....145
   Annex VI
      Regulations for the Prevention of
         Air Pollution from Ships 145
   Record Keeping Requirements ...145, 147
master
   authority 109
   responsibilities.............................109
   review....................................110, 111
mating position 29
mattress 70, 78
maximum
   bending radius 92
   clamping force..............................59
   static thrust 41
medical lock 34
Merchant Shipping
   (Accident Reporting and Investigation)
      Regulations 2005 130
   and Fishing Vessels (Lifting Operations
      and Lifting Equipment)
      Regulations 2006 81
   (Diving Safety) Regulations 2002.........65
   (Health and Safety at Work)
      Regulations 117
   (Means of Access) Regulations 1988...133
   (Pilot Ladders and Hoists)
      Regulations 1999 136
minimum breaking load 86
moonpool............2, 3, 11, 12, 15, 29, 37, 53,
      ....55, 62, 63, 76, 77
   cursor...29
   launch and recovery system 44
mooring 110
   bitts 140
   equipment
      installation 139
      planned maintenance and repairs 140
   fairleads.......................................140
   line 69
      tension 69
   operations 117, 139
   rollers 140
   ropes 140
   safety considerations...................140
   winch 139
   wires 140
multi function tool 42

N

Nautical Institute......................103, 121
navigation and bridge procedures 110
near miss 130
negative buoyancy 40
negatively buoyant.............................26
neutral buoyancy 40
nitrogen 20
   narcosis 20
   oxides (NOx) 146
nitrox 21
NMD Class
   0 101
   1 101
   2 101
   3 101
nominal
   diameter 86
   wire diameter 85
non-conformance 107, 110
   reporting and corrective action 110
non-destructive testing (NDT) sensor 42
normoxic...20
Norwegian Maritime Directorate
   (NMD) 100

O

objective evidence............................108
observation
   ROV 38
      with payload option 38
Occupational Safety and Health (Dock
   Work) Convention, 1979 47
offshore
   construction vessel..........................5
   cranes 47, 84
   lifts 48
   support vessel 1, 6, 7, 8, 10, 11, 12
oil pollution prevention equipment 132
overboarding chutes 61

overside
  A-Frame
    launch and recovery system 44, 46
  camera display..10
  launching system 10
  ROV
    launch and recovery system 46
  trolley
    launch and recovery system 44, 45
oxygen..20
  analyser 19, 26
  and carbon dioxide analysis 31

P

pad inserts 59
pan and tilt assemblies...........................42
passive
  heating system.....................................27
  heave
    compensation..............................81, 82
    mode..82
pedestal 49
permit to work 69
  system 76, 86, 117, 137
    principles 138
    purpose 138
personal protective equipment...86, 90, 132
personnel basket transfer 110
piggy back
  chute 58
  reel 58
  straightener 58
pilot
  access arrangement 136
  boarding operations...........................136
  ladder 136
pilotage
  and pilot boarding.............................110
pipe
  aligner 57, 58
  and cable bridges...............................134
  and product deployment 8
  arrestor..90, 91
  bending radius 6
  clamp 57, 59
    arrangement...............................56, 59
  collar 58
  departure angle....................................59
  end 60, 90, 91
    connection 56
    termination 53
  flexible 6, 54
  lay
    barge 54
    initiation 53
    method 54
    operation 53
    ramp 56
  penetrations 19
  reeling operation 89
pipe lay
  analysis.....................................89, 92, 93
  deployment operation 89
  initiation 93
  operations 89, 90, 93, 95, 110

  rigid and flexible 6
  survey.......................................89, 92, 93
  system 6, 90
  vessel 5
Pipe Lay Vessel 53, 78, 89, 90, 92, 93, 95
  Classification Society design principles.53
  operational design considerations 53
pipeline 37, 54
  analysis..58
  damage..78
  end 59
    termination 93
  flexible 54
  initiation 92
  inspection......................................37, 74
  installation 78
  route 78, 79, 93
  stalk 91
  survey 79
  testing 70
  under pressure 70
Pipe Line End Terminations (PLETs) 60
pitch 95
  and roll control 41
PLET handling frame.................56, 57, 60
ploughing..37
portable ladders 144
Port Facility Security
  Assessment (PFSA) 113
  Officer (PFSO)...................................113
  Plan (PFSP) 113
position
  keeping system 40, 41
  reference system 95, 96
  transducers..40
positive buoyancy....................................41
positively buoyant 26
post
  burial survey.......................................93
  trench survey......................................93
potable
  and hot water supply.....................20, 22
  water 110
power management system 95, 96, 100
pre-dive checklist....................................76
pre-lay survey 92
preparations for spooling........................91
principles of health and safety........117, 118
procedures and working environment....119
product 6
  coating 59
propeller................................6, 53, 95, 100
  and rudder
    combination 8
    configuration 8
  controllable pitch 8
  fixed pitch 8
propulsion
  systems (thrusters).........................40, 41
  unit 2, 53
protection measures..............................119
prototype or development vehicle............38
proving trials 102
provision and use of work equipment....117
  Regulations (PUWER 98)....................143
pull head 91

  assembly..91
  connection assembly...........................91

R

RadaScan 96, 99
radius inserts 60
ram cylinder luffing system.....................51
ramp 91, 93
  arrangement 56
  fleeting..89
    warning beacon 90
  fleeting and elevation system 56
  lay system...92
  system 53
Recognised Security Organisation
  (RSO) 113, 115
Record Keeping Requirements
  (Environmental Protection Act
  1990) 145, 149
reel 53, 55, 90, 93
  lay 54
    installation methodology..................55
reeled pipe operation 89
remote manipulation 42
removal and scour surveys 66
reporting accidents and non-
  conformities 110
response amplitude operators 48
retractable
  azimuth thruster...................................7
  thruster 7
rigid product 93
rigid steel pipe 54
riser 54
risk
  analysis matrix..................................137
  assessment ..65, 67, 74, 76, 77, 83, 86, 90
  and hazard identification..........117, 137
roller
  box 91, 93
  fairleads..59
route survey..92
ROV 8, 11, 37, 95
  auto heading mode.............................77
  buoyancy...76
  classification 38
  components and sub systems................40
  control station.....................................43
  control system 38
  crane operations 5
  dive plan 76
  excursion umbilical 74
  frame 40
  handling systems 3
  hangar 11, 37
  heading 44
  integrity check.....................................76
  launch...76
    and deployment system 37
    and recovery operation 44
    and recovery system 44
    system 10, 11
  launch and recovery 73, 75
  main umbilical 74
  management and control system 42

negatively buoyant 40
observation class unit 38
operational planning 73, 74
operations 2, 3, 8, 10, 37, 73, 95,
....98, 110
   and environmental conditions.....73, 74
   in the vicinity of divers.................73, 77
   in the vicinity of offshore
     installations................................73, 78
   in the vicinity of pipelines 73, 78
personnel 11
pilot 42, 74, 75, 76, 77
   console display 44
   pitch 44
positively buoyant 40
recovery...76
roll 44
supervisor..76
systems..15
types 38
umbilical...45
workclass unit 38
rudder..6, 100
and thruster control 9
Rules for Certification of Diving
   Systems 16

S

safe
   access
     and slips, trips and falls 117, 133
   haven..71, 72
   navigation 8
   working environment........107, 117, 132
   working limits......................................67
   working load 48, 83, 86
safety
   and environmental protection
     policy.................................108, 109, 111
   area inspections 117, 130
   committee....................................128, 129
   committees and safety meetings..........117
   interlock...................................19, 20, 23
     system 35
   leadership 117, 122
   management
     audit 108
     Certificate 108, 109, 110
     system 107, 108, 119, 132
   meeting..129
   officer....................117, 118, 122, 128, 132
   of personnel 89
   representative 117, 128, 129, 132
Safety of Life at Sea Chapter V:
   Safety of Navigation 136
sandbagging of pipelines..........................66
sanitary system 19
saturation
   control 17, 68
   room 18, 31, 32
   diver 77
   dive system..23
   Classification Society Design
     Principles 16
     components 17

operational design considerations.......17
seabed
   sediment 75
   survey.......................................37, 74, 78
Seaeye Tiger observation class vehicle...38, 39
security 10
   assessment 114
   plan 114
sediment removal or reduction 153
self-righting test.......................................33
semi-submersible.....................................68
sequential method 153
sheave
   and main winch diameter....................84
   arrangement...85
   fleet angle...85
ship
   security
     Assessment (SSA) 113, 114, 115
     Certificate (SSC) 114
     Officer (SSO).................113, 114, 115
     Plan (SSP).......................113, 114, 115
shipboard
   cranes..47
   incinerators 147
   management................................108, 109
   Oil Pollution Emergency Plan
     (SOPEP)..145
   safety 117
     organisation 117, 128
shore based management and the master..128
simultaneous
   diving operations..................................77
   operations 66, 74
skid arrangement.....................................46
slack turn...87
S-lay 54
   installation methodology.....................54
   methodology 55
slew bearing...49
slewing angle 82
slips, trips and falls 135
SMS..108, 109, 110, 111
snagging 68
snap-back zone 140
soft umbilical 45
SOLAS.................33, 107, 113, 115, 136
sonar 38
   scanning...39
splash zone 44, 82
   mode 81, 82
spooling 86, 89, 90
   arm 60
   operations 90, 91
spreader beam 84
squeeze pressure...........................58, 61, 91
stability 5, 48, 53, 56, 81, 89, 90
   test 33
stalk 91
standby diver...25
   umbilical...68
stern
   mounted azimuth thruster.....................7
   roller 37
   tunnel thruster 8
still camera..41, 42

stinger..53, 55
straightener 55, 57, 58, 93
   track 58
straightening trials.............................58, 92
strength test...33
structure 37
subsea
   installation construction support..........37
   lift 48, 84
   lifting operations..................................70
subsea installation construction support ..74
sub surface visibility 74, 75
sulphur oxides (SOx)............................146
surge...95
survey...8
   and tank inspection............................110
   of the installation onboard 47
   operations 10, 98, 110
   personnel 11
   support..1
survival bag 28
sway 95
SWL...82

T

taut wire
   arrangement...98
   system..96, 98
telecommunications support 37, 74
tensioner 53, 54, 55, 56, 57, 62, 90,
....91, 92, 93
   linear 58
   pads 59
   tracks 58
tension monitoring system 93
tether management system 39, 42, 43, 76
The Merchant Shipping and Fishing
   Vessels (Provision and Use of Work
   Equipment) Regulations 2006.............143
thermal protection..................................28
through-water communication system.....28
thruster 2, 3, 6, 53, 76
   efficiency...40
   horizontally mounted..........................39
   vertically mounted 39
tie-in arrangement..................................91
TMS
   arrangement...45
   frame 76
   garage..43
tool box talks...90
tooling skid 43
topside installations.................................47
touch down monitoring 93
towed
   and bottom crawling vehicle................38
   array system...37
   vehicle...38
towing..34
tracked tensioner arrangement................59
training 110, 152
   and certification for DP operations 103
transfer
   lock 17, 23
   of waste 148

under pressure (TUP) system 18, 23
transponder...99
    beacon..97
transverse
    thrust8
        effect 8
    trolley arrangement 29
trunking...19
tugger..5, 91
    winches 5
tunnel thruster 7, 100
twin propeller vessels 8
types of
    component gases 20
    mixed gases 21

U

umbilical 54, 55, 58, 76
    and lift wire winch 44
    attachment lug 25
    cable 43
    management.......................65, 66, 68, 74
    tie-ins..70
    winch 45
under deck carousels.................................60

V

variable ballast 41
vertical
    ladders and external stairways.............134
    lay 54
        installation methodology...................55
        system 53
        system tower 60, 62
    pipe lay operations 89
    reference unit (VRU) 100
    (tilt) adjustment 41
visual monitoring 32

W

walkways and working decks 134
waste
    management.......................................110
    Management Controller.....................148
    producers 147
water
    intakes 78
    jetting tool 42
    salinity
        and density.......................................75
        and temperature 74
watertight integrity...................................84
winch drum groove radius.......................84
windlass...139
wind sensor 100
wire
    arrangement
        continuous bend 85
        double bend 85
        reverse bend 85
        single bend 85
    sheave...85
witness statement 131
Work at Height Regulations 2005 143
workclass
    ROV frame 40
    vehicle..38
working at height 90, 117, 143
working radii..51

Y

yaw..95

Z

Z-drive 7